WEAPONS OF MASS DESTRUCTION

The New Face of Warfare

WEAPONS OF MASS DESTRUCTION

The New Face of Warfare

Chitra Lele

Foreword by

James J.F. Forest, Ph.D.,
University of Massachusetts Lowell,
Former Director of Terrorism Studies at
the United States Military Academy,
Professor, Consultant and Renowned Author

PENTAGON PRESS

Weapons of Mass Destruction: The New Face of Warfare / ***Chitra Lele***

ISBN 978-81-8274-709-8

First Published in 2013

Published by

PENTAGON PRESS
206, Peacock Lane, Shahpur Jat,
New Delhi-110049
Phones: 011-64706243, 26491568
Telefax: 011-26490600
email: rajan@pentagonpress.in
website: www.pentagonpress.in

Printed at Aegean Offset Printers, Greater Noida, U.P.

Dedicated to a very
special individual,
my mother, Mrs. Asha Lele,
who has taught me the
meaning of peace.

Also dedicated to the millions of
brave hearts working towards
a weapons-free world.

WMD has two full forms: one stands for the standard full form—Weapons of Mass Destruction, and the other stands for the utopian full form:

Wipe Out
Mass
Destruction.

—*Chitra Lele*

"The more you sweat in peacetime,
the less you bleed during war."

—*Chinese Proverb*

Contents

Foreword

We live in interesting times. With the current global inventory of chemical, biological, radiological and nuclear (CBRN) weapons—often referred to as weapons of mass destruction (WMD)—mankind possesses an incredibly powerful technical capacity to eradicate the entire human race. Further, dozens of countries worldwide continue to invest in developing new kinds of these weapons. Historically, these weapons have been developed (and on some occasions used) by states as part of their military arsenals. The current international system—with its alliances, collective security agreements, reciprocity, and so forth—imposes some constraints on what a state can do with its WMD, but this fact gives us little comfort. A more worrying fear is the day in which a sophisticated WMD lands in the hands of an unconstrained non-state actor intending to inflict massive casualties against its perceived enemies without regard for any consequences.

With this backdrop, a public debate rages about the necessity of these weapons, or lack thereof, for ensuring our security. While recognizing the valuable and legitimate peaceful nature of many CBRN technologies used in the public and private sector, we fear the dual-use potential for the same technologies to be used for developing weapons. This fear is underscored by the international community's recent political and economic sanctions against Iran in response to their overt and secret nuclear programs. In recent decades, other countries have crossed the nuclear threshold—North Korea, India, Pakistan—against the wishes of the international community, and the overarching concern is that more countries will do so in the future. And beyond the nuclear dimension of WMD, we see increasingly powerful breakthroughs in biotechnology and other advanced sciences that also have dual-use potential. The bottom line for the WMD threat that we all face is this: as more CBRN weapons are developed, the more likely it is that they will someday be used.

It is incumbent upon all responsible governments and citizens to reflect on the nature of WMD and its future. As well, millions of professionals in security, law enforcement and first response agencies—from the intelligence and military services to local public health officials, police and fire departments—must be prepared to respond effectively if and when a WMD attack (or accident) occurs. There is also a need for greater public awareness and education. To these ends, the nationally-acclaimed prolific author Chitra Lele has authored an important book, *Weapons of Mass Destruction: The New Face of Warfare,* which synthesizes and integrates an impressive amount of information in way that a general readership can understand. The young author has compiled an impressive collection of technical details and analysis on various kinds of CBRN weapons, making this volume a useful tool for college and university courses, professional development seminars, and other venues in which WMD are studied.

In sum, this book offers an important contribution to the global dialogue on weapons of mass destruction. I agree with those who argue that the world would be a better place for our children if these weapons did not exist. But they do exist, and we must learn all we can about them. We must also work to harness their destructive potential and do all we can to prevent the kinds of devastating attacks that we envision in our darkest of dreams.

James J.F. Forest, Ph.D.,

University of Massachusetts Lowell,
Former Director of Terrorism Studies at
the United States Military Academy,
Professor, Consultant and Renowned Author

Introduction

The Weapons of Mass Destruction (WMD) is a much debated concept for long. However, there is a need to address this issue in the 21st century bit differently. Particularly, post-9/11 with the new face of warfare becoming more obvious, the issues related to WMD need to be debated afresh under the changed geopolitical landscape. It is important to address the issues related to WMD at scientific, technical and political levels. *Weapons of Mass Destruction: The New Face of Warfare*, this comprehensive book by the young author, Chitra Lele, presents a holistic approach to this issue and presents an integrated view on the subject.

Interestingly, in the post-9/11 era and particularly after the United State's invasion of Iraq in 2003, there has been a perceptible shift in attitude of some in the world in regards to the threat from WMD. Actually the absence of WMD in Iraq has made few to believe that this is an over exaggerated threat. However, it is important to appreciate that the "intelligence failure" in Iraq no way diminishes the subsistence of this threat. Particularly, the 2011-12 Jasmine Revolution in West Asia and adjoining regions goes on to prove the existence of WMD in Libya, Syria and probably in Egypt too. Even today the remains of Saddam Hussein's chemical weapons stockpile are still present in Iraq (incidentally, Britain is planning to help the Iraqi government in its disposal by 2012). States like North Korea and Iran are the most disputed WMD possessors of the day. Presently, the threat exists in all domains of WMD, viz., nuclear, biological and chemical arenas. It is important to note that in this era of asymmetric warfare, the derivative of such weapons extends this threat further to the arenas of radiological and agricultural terrorism.

Luckily, no major successful WMD attack has taken place since 1980s when the chemical weapons were used by Saddam Hussein during the Iran-Iraq conflict. The success for this probably goes to the United Nation's disarmament and arms control regime. However, the relevance of such mechanism weakens in the scenario of state versus non-state conflict where the non-state actor is uncontrolled and is not bound by any global norms.

This book could be viewed as a comprehensive source on this subject under the changed geopolitical and geostrategic settings of the 21st century. The book covers a wide canvas from history to character of such weapons and disarmament to technology. It also makes the reader aware about the probable WMD future by developing few futuristic scenarios.

The book traces very many details of individual WMD agents which hold the potential for misuse. Particularly, in the realm of biological and chemical weapons, such detailed analysis is important for better understanding of the subject not only for the purposes of appreciating the threat but also for developing the protective and defensive measures.

Nuclear weapons are unarguably the most important elements of WMD spectrum. These weapons have relevance both from strategic perspective as well as nuclear/radiological terrorism perspective. Its deterrence potential still remains the undisputed aspect of global affairs. Over the years much has been debated on the geopolitical aspects of this issue. This book takes this debate further and particularity discusses various warhead and other technical aspects of this issue and also provides the historical backdrop of evolution of these weapons.

This book presents various historical, political and technical aspects of WMD. It examines the scope and limitations of WMD in the background of new face of warfare. The author of this book has presented various features of WMD in a very lucid form and I am hopeful that the book will be well received by the readers.

Wing Commander Ajey Lele, Ph.D.,
WMD and Strategic Technologies Expert,
Author of *Bio-Weapons: The Genie in the Bottle,* and
Research Fellow, Institute for Defence Studies and Analyses,
New Delhi, India

Acknowledgments

I have always felt that writing acknowledgments is the most humbling aspect of writing a book. The process of writing a book involves the support and guidance of many people. I am pleased to express my gratitude to those who have helped me either in a direct or indirect way while writing this book. Thank you all for believing in my book and in me.

First and foremost, special thanks go to my parents, Mrs. Asha Lele and Mr. Lele. Whenever my spirits lagged, their constant support and faith helped me to regain my intellectual energy. I dedicate this book to them. It is through their teachings and blessings that I have grown as an individual. They are an oasis of wisdom, which has added to my intellectual maturity level in more ways than one.

I give my sincere gratitude to God for being my mentor and helping me along the way in so many mysterious ways.

Prof. James J.F. Forest has been a priceless source of valuable inputs. I thank James Sir profusely for his steady encouragement and for writing the scholarly Foreword for my book.

I am heartily thankful to Wing Commander Ajey Lele, whose valuable guidance from the initial to the final stage enabled me to complete the book successfully. Thanks to Ajey Sir for taking the time to write a scholarly Introduction to my book.

Without the comprehensive research studies conducted by several premier organizations, sharing of vital research findings, valuable actionable insights and latest statistics would have not been possible. Many thanks to these organizations for their relentless efforts in keeping us abreast of the latest developments in the dynamics of Weapons of Mass Destruction.

I gratefully acknowledge the members of the Brainlines Team (www.brainlines.net) for their incredibly creative work on the figures. Special thanks to Wikipedia.org, the online comprehensive encyclopedia, which provided us the base images to adapt for the book.

I am grateful to all the experts with whom I had illuminating discussions. It is because of these thorough professionals that we get to learn interesting theories, developments and facts about weapons of mass destruction.

I want to gratefully acknowledge the readers for committing their valuable time to reading this book.

Last but not least, I am extremely thankful to my coveted publishing house, Pentagon Press, New Delhi. Rajan Arya Sir, the CEO of Pentagon Press, has been very kind in giving me this wonderful opportunity to write an academic book on such a critical topic of global security concern. Rajan Sir's useful tips served as a sounding board during the writing process of this book. The steadfast support and ceaseless cooperation from the entire Pentagon Press Team is greatly acknowledged.

Chitra Lele

Preface

Intended as a starting point for understanding weapons of mass destruction (WMD), the book *Weapons of Mass Destruction: The New Face of Warfare,* is a comprehensive resource on the most deadliest weapons of all times. With the current evolution of potential WMD threats, it is important to understand these threats and their implications in order to deal with them effectively. WMD have come a long way from poison-tipped arrows and diseased animal carcasses to high-tech, genetically modified organisms and aerosolized release of contagious viruses. In these days of a constant threat from unconventional weapons, my book takes you safely through the labyrinth of weapons—their evolution, specifications, underlying design, features and many more aspects. It cuts through the technical jargon to maximize the reader's understanding of the subject so that the global community can work out strategies for clearing the cloud of chaos and panic, developing multilateral control and improved intelligence, and defusing the threat of WMD.

Through the various sections of the book, I have dealt with the *how, who, when, where, why, what* and *what next* of this critical topic of global security concern in an easy-to-understand manner. This book is an attempt to cover the many faces and dimensions of WMD. It provides useful resources, facts and information, which will enrich the bookshelves of home, educational and research libraries.

The book deals with:

- Historical Context of WMD
- Major Types of WMD
- Facts and Scientific Information of WMD
- Future Scenarios of WMD
- Treaties, Protocols and Conventions of WMD
- And much more ...

This book is divided in five parts, each with its own unique focus. The reader's journey begins with the history of WMD and goes on all the way to the modern-day WMD. Part I covers the foundation topics like the history, academic definitions, types, treaties, protocols, and the stand of countries towards WMD. The last section in Part I discusses the new types and future scenarios of WMD like economic warfare, nanotechnology, etc. Part II deals with various dimensions of the deadliest weapons of WMD spectrum, Nuclear weapons. The chapters in this part deal with a gamut of topics, right from types and treaties to weapons design and nuclear strategy, and from symptoms and effects to responses, treatment and more. Part II delves further into nuclear weapons—it discusses in detail the nuclear weapons of major nuclear powers like the United States, Russia, United Kingdom and France. Part III deals with Radiological weapons, right from the types and delivery systems to effects and responses. Part IV deals with the domain of Biological weapons. It discusses the various biological agents and their details in terms of overview, delivery systems, potential as a biological weapon and bioterrorism threat, and symptoms and management. Part V focuses on the dynamics of Chemical weapons, right from basics to delivery systems and from the various chemical warfare agents to their potential as weapons of mass destruction. Appendix A provides a snapshot of the various learning material and useful resources available for further reading and research. Appendix B shares information about research institutions, private organizations, government bodies and international agencies that are working relentlessly towards realizing the vision of a world free of WMD. Appendix C contains more academic definitions of WMD.

The audience for this book is everyone who is affected by war and terrorism, either directly or indirectly, and wants to join the bandwagon of anti-war activism, i.e., You. Through this book, I want to provide some food for thought to spark up the intellectual independence within each one of us to think of ways of building a weapons-free world.

This book is written with a purpose to:

1. Make the common man aware of the various dimensions of WMD.
2. Trigger constructive, spirited discussions in the classroom and in the academia among the analysts, concerned citizens, governance specialists, policymakers, investigative journalists, social activists, public health officials, community leaders, research students and heroes of anti-WMD.
3. Encourage individuals, expert groups and institutions to analyze WMD scenarios, both current and future, and provide concrete inputs towards curbing them.

All the tools needed to facilitate quick understanding are provided—spotlight and a word of caution sections, inline definitions of technical terms, latest statistics, lists of resources for further study and research, questions for reflection, figures and tables that clarify concepts, and much more. It is my earnest attempt that my book contributes to a better understanding of the various dimensions of WMD, and to a better and doable agenda against them.

Chitra Lele

List of Figures and Tables

List of Figures

List of Tables

PART I

Introduction—The Lethal Legacy of Weapons of Mass Destruction

1

Weapons of Mass Destruction—Warfare Redefined

Advancements in modern technology have opened the door to more destructive weapons—*Weapons of Mass Destruction* (WMD). These advancements have increased the destructive power of WMD manifold; devastation caused by WMD is far more powerful than the destruction that can be achieved by conventional means.

The term WMD first appeared in December 1937 in an address given by the Archbishop of Canterbury, William Cosmo Gordon Lang. Its modern usage dates to 1945 with the insertion of the words "weapons adaptable to mass destruction" in a document signed by the then American President Harry Truman. An early use of the exact phrase in an international treaty was in the Outer Space Treaty of 1967. The term WMD may have been in use only since the past few years, but these weapons have been long in use, right from ancient arrows to modern marvels.

As per Allison Macfarlane, a specialist on WMD, nuclear weapons are far more destructive and certain as killers than either biological or chemical weapons. During the Cold War, the aspect that scared people the most was the threat of a nuclear war, particularly an all-out nuclear war that would wipe out the existence of the world. Compared to chemical weapons, biological weapons are far more deadly. Nuclear weapons are the most difficult and expensive to develop. In sum, nuclear, biological and chemical weapons "are fundamentally different in terms of lethality, in the area they cover and over time; in the availability of measures that can protect against them" (Morrison and Tsipis, 2003).

The Cold War has left an indelible mark on the dynamics of arms control and international security, and has a long-lasting impact on WMD and their developments. As biological and chemical weapons can be prepared, packaged, delivered, and set off with ease, they are much more likely to be a weapon of choice for terrorists and rogue states, and therefore since the end of the Cold War, the main concern regarding WMD has been proliferation.

The theories, experiments and discoveries of eminent scientists of the 20th century provided the basis of nuclear research and development. The modern history of biological warfare started in 1918. The widespread use of chemical warfare in modern times began in 1915. The history of radioactive weapons may be traced to a 1943 memo from James Bryant Conant, Arthur Holly Compton and Harold Urey to Brigadier General Leslie Groves of the Manhattan Project.

Essential actions for WMD include detection, prevention, preparation, response, mitigation, treatment and reassessment. Controlling the spread of WMD through these actions depends on how easy it is to detect the weapons programs carried out across the globe.

Figure 1.1: The Lethal Cloud of WMD

The use or threatened use of WMD has significantly influenced the 21st century and will continue to impact international security in the foreseeable future too. With the help of WMD, even small groups of individuals have the ability to cause enormous damage. Their very existence is a real threat to the survival of civilization itself. Over the decades as WMD arsenal has increased, the globe has continued to quake with incessant fear and is constantly being orbited by the lethal cloud of WMD.

WMD are used for not only causing destruction of life and property, but also as a means of causing tremendous psychological trauma, achieving political goals and creating a wave of mass panic. The aftermath is just as terrifying as the actual WMD attack. The victims suffer from post-traumatic anxiety disorders and stress-related emotional or physical symptoms. The immediate impact of these weapons reaches much further than the actual site of attack.

From the days of infected bodies hurled at the enemies to the age of Bioterrorism agents, WMD have come a long; today they are products of ultra-modern science. They also have a considerable impact on popular culture. So much so that movies, books, science fiction, media stories and TV shows have them as their central theme.

Though WMD has become a part of the global weapons lexicon, for the common man it is still an enigma. In this chapter as well as in the rest of the book, we will discuss everything related to WMD in order to enhance our understanding of the same.

1.1 History of Nuclear Weapons—From Path-Breaking Theories to Modern-Day Realities

The history of nuclear weapons began in 1896, when Antoine Henri Becquerel discovered radioactivity in uranium. Then in 1902 Marie and Pierre Curie isolated the radioactive metal called radium; an ore of uranium that could produce large amounts of radioactivity. Three years later, Albert Einstein published his path-breaking theory of relativity, wherein he stated that transformation of mass into energy would "liberate" huge amounts of energy. Further experiments by James Chadwik, Ernest Rutherford, John Cockcroft and Ernest Walton gave a new hope to the scientific community that there are elements containing tremendous amounts of unseen energy and this energy can be harnessed. This theory needed an action-oriented approach, which was taken by Niels Bohr and Ernest Rutherford, who concluded that it was the nucleus that had to be broken, disintegrated, and "exploded", if atomic energy was to be liberated.

The invention of nuclear weapons was the result of the political and scientific progress made in the 1930s. In the 1930s, there was enormous progress in nuclear research. In 1934, the Italian physicist, Enrico Fermi, discovered nuclear fission by bombarding uranium with protons. He disintegrated heavy atoms by spraying them with neutrons. In 1938, a similar experiment with uranium was carried out by Otto Hahn and Fritz Strassman; they were able to split an atom and achieve nuclear fission. These scientists had managed to put Einstein's theory into action by way of their revolutionary experiments.

After about 8 months, Einstein wrote a letter to the then American President, Franklin Roosevelt, recommending that atom bombs be made. Thereafter, the Manhattan Project was initiated; a massive undertaking to develop the first fission weapons jointly by the United States, Britain and Canada during the World War II. Top scientific minds of those times were brought together for producing fission-based explosive devices.

The world's first nuclear test, codenamed 'Trinity,' was conducted on 16th July 1945 by the United States in New Mexico; a result of a top-secret initiative, the Manhattan Project. And three weeks after the test, on 6th August 1945, the Enola Gay, an American airplane, dropped the first atomic bomb ever used in warfare on Hiroshima, Japan. On 9th August 1945, a second atomic bomb was dropped on the Japanese city of Nagasaki. These events brought a close to the World War II and marked the beginning of the Cold War between the United States and the Soviet Union. Other countries followed: The Soviet Union was the next country to test a bomb on 29th August 1949. Britain's first test was on 3rd October 1952. France's on 3rd December 1960. China's on 16th October 1964. India's on 18th May 1974. For more than four decades, the two superpowers, the United States and the Soviet Union threatened each other with nuclear weapons. This time period is referred to as the *Cold War.*

During this same time period, scientists were also working on developing bombs based on the principles of fusion—mass could be converted into vast amounts of energy through fusion by taking atoms of a light element such as hydrogen and fusing them. Edward Teller, also known as the father of the hydrogen bomb, researched and developed relentlessly, along with his colleagues' ideas and principles, and produced the first workable design for a megaton-range hydrogen bomb in 1951. On 1st November 1952, the Teller-Ulam configuration, codenamed "Ivy Mike," a two-stage thermonuclear device, was tested in Enewetak Atoll in the Pacific; this is when the "thermonuclear age" began. The first deployable Teller-Ulam weapon in the United Sates was developed in 1954. Thermonuclear weapons can produce much larger explosions than fission weapons; their enormous explosive power is 25 times more powerful than the first nuclear bombs.

Since 1945 there have been only two instances of the actual usage of nuclear weapons in warfare. More than 2,000 tests have been carried out across the world in sites like Pokhran (India), Enewetak Atoll (United States), Maralinga (United Kingdom), etc., in all types of environments—above ground, underground and underwater.

1.2 History of Biological Weapons—From Decomposed Bodies to Deadly Agents

The use of biological weapons dates back as far as 300-400 B.C.—Scythian archers infected their arrows by dipping them in decomposing bodies or in blood mixed with manure. According to Herodotus, a Greek historian of the fifth century B.C., Scythians used a mix of decomposed bodies of venomous adders, human blood and animal feces, and placed it into sealed vessels and buried this mixture until it was sufficiently putrefied. We can call them as pioneers of primitive biological weapons.

During the siege of Athens by the Spartans, a deadly epidemic broke out which killed thousands of Athenians, and eventually resulted in the victory of the Spartans. According to the famous Greek author and historian, Thucydides, this victory was achieved by poisoning wells. It has never been known what organism caused the epidemic, but speculations about it being the Ebola virus are made even today.

In 184 B.C., the great military commander of the Carthaginian Army, Hannibal, used biological warfare to win the naval battle over Eumenes II of Pergamon. Jars filled with venomous snakes were hurled at the enemy vessels.

In the 14th century, the Tartar army used a combination of psychological warfare and biowarfare. The outbreak of the bubonic plague helped them to throw over the plague-infected bodies in the enemy's territory in order to initiate an epidemic, which resulted in mass exodus. The Black Death (plague) killed millions of people, approximately one half of the population, throughout Europe in the 14th century. During the Middle Ages, victims of the bubonic plague were used for biological attacks.

In 1650, the Polish fired hollow canister shells filled with the saliva of rabid dogs towards their adversaries.

During the 19th century, it is believed that the trade of many items including blankets by Richard Henry Dana with Mexicans and Russians, who had established outposts on the northern side of the San Francisco Bay, caused the California plague epidemic. It is possible that the blankets were the source of the plague.

Before the 20th century, there were three major ways of using biological agents—use of biologically inoculated fabrics, use of microorganisms, toxins or animals, living or dead, in a weapon system, and deliberate poisoning of food and water with infectious material. For example, during the 1785 siege of La Calle, Tunisian forces flung diseased clothing into the city.

In the 20th century, during the World War I, Germany is said to have used anthrax and glanders in an effort to destroy animals used during the war. Germany pursued an ambitious biological warfare program throughout the war period, which triggered a chain reaction that made countries like the United States, Canada and United Kingdom launch their own biological weapons programs.

Japan ran a research unit in Manchuria during the World War II where prisoners were exposed to agents such as plague and anthrax. There are reports stating that a number of other countries were thought to have considered the idea of using biological weapons during the World War II, but there is no documented evidence of any other use of biological weapons in the World War II. The Japanese use of biological warfare agents against the Chinese led to major powers like the United States, United Kingdom and Soviet Union undertaking R&D programs.

SPOTLIGHT

Diseases that are considered as major biological weapons in modern times have been known since ancient times.

In recent years, new and deadly agents like Ebola have begun to dominate the biological weapons terrain. During the past century, several tens of thousands of innocent people have died due to the release of pathogens or toxins.

1.3 History of Chemical Weapons—The Poor Man's Nuke

Crude forms of chemical weapons have been employed in ancient times, for example, in the late Stone Age hunters in Southern Africa used arrows tipped with poison. In ancient times, during the siege of Kirrha, Solon of Athens is said to have used hellebore roots to poison the water supplies of adversaries. Ancient Greek myths about Hercules poisoning his arrows with the venom of the Hydra Monster are the earliest references to chemical weapons in western literature. Even the ancient Indian epics of Mahabharata and Ramayana contain references to chemical warfare. Scientists have discovered Chinese writings containing hundreds of ways to create chemicals that were used in war.

Drinking water wells were poisoned with rye ergot by Assyrians and Persians during the 5th and 6th centuries B.C. In the Middle Ages and Renaissance period, barrels of blinding quicklime were catapulted by the English fleet on French vessels (middle of the 13th century). In another instance, grenades, bombs and rags containing arsenic were fired by the defenders of Belgrade against the Turks in 1456. Even Genghis Khan's Mongolian forces employed chemical warfare when they catapulted burning sulfur and pitch into cities they besieged.

The modern era of chemical weapons began during the World War I, when Germany used chlorine gas from cylinders to eliminate the Allied troops; this toxic cloud killed almost 5,000 Algerian and French troops. The French put phosgene into a standard shell and this became the common method of delivery. More than two dozen different chemical agents were used during the World War I. After the World War I, chemical agents were occasionally used to suppress rebellions or subdue people.

Despite the 1925 Geneva Protocol banning the use of these weapons, after the World War I, chemical weapons were used on several occasions—Italy in Ethiopia (1935-36), Japanese used chemical warfare against China in 1930.

Although no chemical weapons were used in the World War II except for Japan, many countries on both sides had huge stockpiles of them.

In the post-war years, despite the arms race between nations, chemical weapons were used on a large scale only twice during the remainder of the 20th century—Egypt used mustard gas in Yemen (1963), Iraq used mustard gas, cyanide and tabun against the Iranian troops (1980-88).

Today, thousands of innocent people remain chronically ill as a result of exposure to chemical agents. Chemical weapons remain the 'poor man's nuke' as they provide a low cost and low technology alternative.

1.4 Definition—Much more than an Acronym

What is WMD? The term "WMD" does not have a universal meaning. However, different nations, groups and associations have developed their own definitions. In common parlance, WMD is a weapon with the capacity to inflict massive-scale death and destruction. But when it comes to disarmament, security considerations, international policies and international conflicts, this term takes on a widespread perception and its meaning changes from nation to nation, from policymaker to policymaker, and from speaker to speaker. There is no one authoritative definition present. Depending on the definition adopted the framework and scope of combating WMD programs and the implications of their use to guide policy change substantially.

Some experts are constantly coining radical definitions while some pundits prefer to drop the term altogether. Dropping the term is not possible as it an integral part of the world weapons landscape, arms control treaties, political discourse and military doctrines. The growing importance of this term is testified by its frequent usage in academic literature and press articles. For example, the *New York Times* (New York Times is a leading American daily newspaper) used the term at least once a year from 1945 to 2005. Because of its prolific use, the *American Dialect Society* (American Dialect Society is dedicated to the study of the English language in North America and in other parts of the world) voted WMD as the word of the year in 2002.

WMD are nuclear (N), biological (B), chemical (C) and radiological (R) weapons; they are collectively referred to as CBRN weapons. The traditional meaning of WMD includes nuclear, biological and chemical weapons (NBC), and in the military context, it refers to atomic, biological and chemical warfare (ABC), whereas the international community prefers to include radiological weapons in the definition too.

Let us gain an understanding of some of the existing definitions.

The UN adopted a standard definition in 1948: [WMD are] ... atomic explosive weapons, radioactive material weapons, lethal chemical and biological weapons, and any weapons developed in the future which have characteristics comparable in destructive effect to those of the atomic bomb or other weapons mentioned above.

According to the Russian military doctrine, WMD means "Weapons used to inflict heavy casualties. They include nuclear, chemical, and bacteriological agents."

These weapons cause death or serious injuries by releasing or disseminating poisonous chemicals, biological agents or toxins. According to the definition defined by the United States law: the term "weapons of mass destruction" means—any weapon that is designed or intended to cause death or serious bodily injury through the release, dissemination, or impact of toxic or poisonous chemicals, or their precursors; any weapon involving a biological agent, toxin or vector.

Let me add another perspective—in the wake of the World Trade Center attacks of 2001, the term WMD has come to include any means capable of inflicting mass casualties. These attacks have made it loud and clear that there are groups with considerable financial and human resources, and a strong will to inflict mass destruction.

Other documents expand the definition of WMD to also include radiological or conventional weapons. The United States military refers to WMD as "Weapons that are capable of a high order of destruction and/or of being used in such a manner as to destroy large numbers of people. Weapons of mass destruction can be high explosives or nuclear, biological, chemical, and radiological weapons, but exclude the means of transporting or propelling the weapon where such means is a separable and divisible part of the weapon."

In a recent definition (2005), the United States Strategic Command (USSTRATCOM) defines WMD "as weapons—nuclear, biological, chemical and radiological—and their means of delivery that are capable of a high order of destruction and/or of being used in such a manner as to destroy large numbers of people or cause significant infrastructure damage."

SPOTLIGHT

Despite debate over its appropriateness and its definition, WMD is a term that is going to be around for a long time.

From the above definitions, we can conclude the following key points about WMD:

A. WMD inflict large-scale death and injury in one of the following ways—a disease organism, radioactivity or poisonous chemicals; these are weapons that have a relatively large-scale impact on people, property, and/or infrastructure.
B. WMD is an integral part of the world weapons landscape, arms control treaties, political discourse and military doctrines.
C. WMD is collectively referred to as CBRN weapons.
D. WMD is a term of saliency in disarmament negotiations and international agreements.
E. Depending on the definition adopted the framework and scope of combating WMD programs and the implications of their use to guide policy change substantially.
F. While these weapons have different design features, delivery systems and effects, they all share one aspect—they are indiscriminately lethal.

1.5 The Types—From the Silent Threat to the Devastating Explosion

WMD include nuclear, radiological, biological and chemical devices. They range from the silent threat of a poison gas attack to the devastating nuclear explosion. Let us take a quick scan of the major types of WMD:

A. Nuclear—they are the most serious looming threat that has the potential to kill millions of people. The radioactive materials released from nuclear weapons continue to affect people for many years even after the explosion.
B. Biological—they include toxic materials derived from toxins or organisms such as fungi, bacteria, rickettsiae and viruses found in nature, or they can be artificially manufactured toxic substances too. The main purpose of these agents is to affect human life and health, animals, plants, soil, air and water in a variety of ways. These substances work to kill or incapacitate the host.
C. Chemical—they are made from poisonous gases and liquids that produce physical and physiological effects on the target. They choke their victims, poison their blood, blister their skin or disrupt their nervous system.
D. Radiological—they use conventional high explosives to disperse radioactive material like radioactive medical waste over the target, for example, to force evacuation of a major city or to contaminate agricultural land. When high explosives are used to disperse the radioactive material, radiological weapons are called "dirty bombs."

Figure 1.2: WMD Hazard Symbols

Hazard symbols are warning signs about hazardous materials, conditions or locations.

Each of these weapons has an entire section dedicated to them, where we will delve further into the various types, design features, effects and much more.

1.6 Protocols, Treaties—Nonproliferation, Bans and More

Several international treaties, protocols, conventions and agreements have been enacted to regulate the use, development, research, spread of technology, and the possession of WMD, although not all countries have signed and ratified them. Some of the main treaties, protocols, conventions and agreements include:

The Treaty on the Nonproliferation of Nuclear Weapons, commonly known as the Nonproliferation Treaty (NPT)—it seeks to inhibit the spread of nuclear weapons and weapons technology to further the goal of disarmament; it provides the basis for international cooperation on stopping the spread of nuclear weapons. It opened for signature in 1968 and entered into force in 1970. A total of 190 parties have joined the Treaty, including the five nuclear weapons states: United States, Russia, United Kingdom, France and China. It has three main "pillars"—nonproliferation, disarmament and peaceful use.

The Seabed Arms Control Treaty or Seabed Treaty—with advances in technology of oceanography, the world was worried that the ocean and the seabed would provide a new environment for warfare; hence the Seabed Treaty was introduced with the intent to prevent the introduction of international conflict and nuclear weapons into new areas and environments. It prohibits the placement of nuclear weapons or weapons of mass destruction on the seabed, the ocean floor, and the subsoil of the seabed. This treaty was opened for signature in Washington, London and Moscow on 11th February 1971 and it entered into force on 18th May 1972.

The Outer Space Treaty—it was signed in January 1967 and entered into force in October of that same year. This multilateral agreement signed and ratified between the U.S., U.S.S.R., and U.K. bans the placement of nuclear weapons or "weapons of mass destruction" in orbit around the Earth, bans installing them on the moon or any other celestial body, and bans the use of the moon or any celestial body for military purposes, including weapons testing of any kind. As of October 2011, 100 countries are states parties to the treaty, while another 26 have signed the treaty but have not completed ratification.

The Geneva Protocol—the horrors of the World War I led to the 1925 Geneva Protocol; it was a response to the use of poisonous gases in the World War I. The protocol was created during the Conference for the Supervision of the International Trade in Arms and Ammunition held by the League of Nations in Geneva in 1925 for the prohibition of asphyxiating, poisonous or other gases, and of bacteriological methods of warfare. It was signed at Geneva on 17th June 1925 and entered into force on 8th February 1928. The Geneva Protocol thus restated the prohibition previously laid down by the Versailles and Washington treaties and added a ban on bacteriological warfare.

The Biological and Toxin Weapons Convention (BWC)—in 1969, after strong protests against the use of herbicides in Vietnam, the United States unilaterally banned the production, development and stockpiling of biological weapons. This led to a series of agreements culminating in the BWC of 1972; the first international disarmament treaty banning an entire category of weapons of mass destruction. It completely bans countries from developing, producing, stockpiling or otherwise acquiring biological agents for use against an enemy.

The Chemical Weapons Convention (CWC)—is an arms control agreement which prohibits the production, use, acquisition, development, stockpiling and indirect or direct transfer of chemical weapons. It is a landmark agreement that has proved its effectiveness for improving international security.

Post-9/11 attacks and subsequent anthrax scares, the top concerns and challenges of international security policy agenda are first, rogue states have or are seeking to possess WMD and second, rogue states provide direct and indirect support to terrorist groups. Decades of arms control negotiations have greatly reduced the number of weapons around the world, but there is no current strategy that guarantees full protection against the risks from WMD. And with the new challenges, traditional arms control strategies and treaties will need to realign their focus with emerging radically different strategic cultures in order to control the flow of WMD, their delivery systems, and the exchange of technical know-how and related materials.

1.7 Countries and WMD—Their Stand, Programs and More

In the case of nuclear weapons, there are five 'declared' nuclear weapons states. They are the United States, United Kingdom, Russia, France and China. 'Declared' means that they have signed the Nuclear Nonproliferation Treaty (NPT) as countries which have tested nuclear weapons; these five are considered to be "nuclear weapons states" (NWS) under the terms of the Nuclear Nonproliferation Treaty (NPT). China, Pakistan and India have exploded nuclear devices in test blasts. The only country to have used a nuclear weapon in war is the United States. Countries with access to nuclear weapons through nuclear sharing agreements include Belgium, Germany, Italy, the Netherlands and Turkey. In only one instance has a nation given up nuclear weapons—South Africa voluntarily gave up nuclear arms in the 1990s. Libya voluntarily renounced its secret nuclear weapons efforts in 2003. Israel is not a member of the Treaty on the Nonproliferation of Nuclear Weapons (NPT) and it is pursuing a policy of strategic ambiguity with regard to its possession—it refuses to officially confirm or deny having a nuclear arsenal or nuclear programs.

Many countries, including signatories of the BWC, currently pursue research into the protection against biological weapons, which is not prohibited by the BWC. Despite the signing of the BWC, it is very difficult to get independent verification on who possess biological weapons. There are several countries that continue to make and study them, for example, Russia and the United States. Nations like Syria and North Korea, among others, are thought to still possess biological weapons. Although some intelligence estimates suggest that various nations possess bioweapons, it is difficult to confirm these estimates as there is very limited open-source information available in this regards.

The United States and Russia have known stockpiles of sarin (sarin is a lethal cholinesterase inhibitor). It is also believed that India, South Korea and Syria, among others, have supplies of various nerve agents. A few countries have pursued the development of radiological weapons, only to abandon these efforts in favor of more practical and effective weapons. By 1953 the USSR developed a radiological warhead for the R-2 ballistic missile, which was retired as nuclear warheads became available. In 1987, Iraq pursued development of radiological weapons, but this was discontinued as Iraq concentrated on nuclear, biological and chemical weapons programs.

A WORD OF CAUTION

Despite the deterrence strategies and arms control treaties, the use of WMD has continued even in recent years.

1.8 The Future Landscape—New Types of WMD

International security is now complex and unpredictable. With gaps in the nonproliferation policies and protocols, and possibilities of deadlier weapons coming down the pike, terrorist attacks like 9/11 will always be looming large over the horizon.

In today's era of globalization, the world operates no more in isolation. Research, knowledge and scientific development have managed to diffuse boundaries successfully. Improvements in international transport and telecommunications have made it possible for easy movement of goods and technical know-how. Therefore, researchers and scientists of any field are able to have access to expertise of multinational teams of specialists in different countries. These developments will certainly have implications on the development of WMD and their proliferation. Embargos and restrictions like export control measures and restrictive sanctions can no longer deter nations to continue with their WMD programs. Such restrictive sanctions and export control measures will be effective in supporting nonproliferation only when there is a high degree of multilateral support.

The term "weapons of mass destruction" as defined by the Commission for Conventional Armaments, in its first resolution, adopted in 1948, "all armaments and armed forces, except atomic weapons and weapons of mass destruction fall within its jurisdiction" and also stated that "weapons of mass destruction should be defined to include atomic explosive weapons, radioactive material weapons, lethal chemical and biological weapons, and any weapons developed in the future which have characteristics comparable in destructive effect to those of the atomic bomb or other weapons mentioned above." Even this definition hints at including future WMD.

As extraordinarily small quantities of biological weapons provide extraordinary lethality; hence biological weapons, both as WMD and as tailored military instruments, could well become the weapon of choice by 2025. And the modern-day well-equipped university or hospital laboratories are quite adequate for the production of many biological warfare agents.

SPOTLIGHT

As it becomes less possible to control potential future WMD, we must control the users of such weapons, particularly those who have the potential to develop these new WMD.

New categories that can be included in the pantheon of WMD of the future are Nanotechnology, Artificial Intelligence, Cyber Weapons like digital warheads (for example, Stuxnet computer worm used to attack Iran's nuclear installations), and Economic Warfare. But again what can be labeled as WMD as far as these new categories is concerned will need in-depth focus and analysis. But with these new weapons, offensive weapons may overpower defensive weapons thereby managing to disturb the modern world's military balance. Nuclear, biological, chemical and radiological weapons are as dangerous to their makers as they are for their intended targets. Moreover, the materials, components and technologies needed to produce these weapons are closely being monitored, which again keeps a tab on their movement. This is not the case with new weapons like economic warfare and cyber espionage. As the world continues to integrate in all spheres and experiences a worldwide knowledge explosion, it will become necessary to restrict individual liberties and freedom even more in order to control the users of such weapons; decrease in personal freedom for an increase in global security. It is necessary to institute standards and norms governing all types of research and development projects and facilities in order to encourage legitimate research efforts and prevent illegitimate developments.

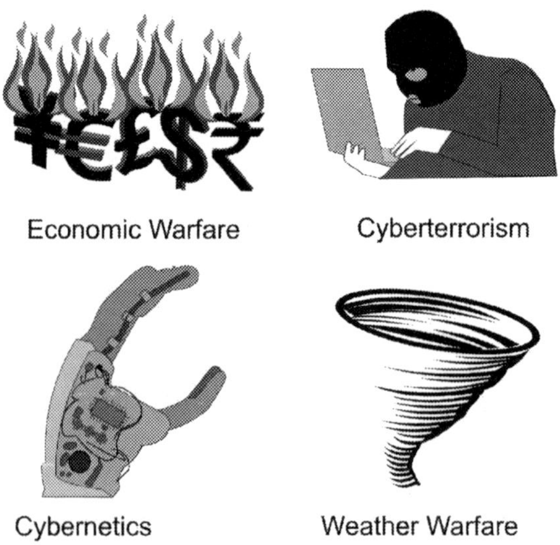

Figure 1.3: New Scenarios of WMD

1.8.1 Economic Warfare, Cyber Warfare and More

Economic warfare is all about destabilizing an economy in order to force a nation into political or military submission; hence crippling the resources of high economic and military value, including the lives of the general population (starvation, bankruptcy). It refers to the use or the threat to use economic means against a country as a part of military operations during wartime. There are several tactics employed like devaluation, blacklisting, suspension of aid, embargoes, etc. These means have been used in the past (right from ancient Greece wars to modern times), but with its modern-day broadened usage, economic warfare now has global level implications. One glaring example of cyber WMD that comes to my mind is Cyberterrorism. With the advent of the Internet, potential users of WMD like terrorist groups have design information and technical expertise at their fingertips. Unlike a virus program that only leads to denial of service, a Cyberterrorism attack is intended to cause physical harm and financial loss, and disrupt life. Cyberterrorism, through a one-click operation, rogue states and terrorists can not only enhance the terror level but also cripple the confidence of the society. For example, by hitting the enter key, they can unlock the valves of a nuclear reactor and cause the same damage as that of a nuclear plant accident. Terrorists can go one step further and can use cyber space to enhance the effectiveness of their conventional means of attacks too. For example, they can strengthen their conventional attack by using cyber warfare tactics to cripple the helpline system so that the victims do not receive timely aid from helpline services. Not every piece of cyber malware should be considered as WMD. Not every control or embargo should be considered as a means to contain WMD. This aspect will again need long-term discussions across the globe to discern the real threats from such emerging attack scenarios.

The rapidly advancing field of biotechnology has the potential to create a much darker biological warfare threat. As progress is made, several new virulent forms of weaponized agents can be developed that could be designed to evade an autoimmune response, cause higher morbidity or mortality rates, or act as a stealth virus that could lie dormant within a victim's body only to be made active by a certain trigger. Organisms with altered characteristics are the "next generation" biological weapons—something referred to as Black Biology. Without vaccines for such designer agents and designer diseases, mass destruction is bound to occur. Such near-term capabilities can definitely give a boost to bioterrorism; they can give terrorists creative but lethal delivery methods. In late 2001, the anthrax letter attacks served as a preview of the future. New weaponized airborne forms are the most dangerous as one has no way of knowing them when one comes in contact with them; one can only hope to be prepared to respond after the fact. Modern techniques like combinatorial chemistry could result in the emergence of new chemical warfare agents, and the harmful materials disposed off from such techniques can in turn give a boost to chemical terrorism, whereby terrorist groups can acquire such materials for their destructive intentions.

A WORD OF CAUTION

What makes these future scenarios even more dangerous is that these new types do not need large facilities or rare raw materials. Moreover, one need not be suicidal to unleash mass destruction through these new weapons.

Another potential source of future WMD is the 21st century technologies of Genetic Engineering, Genome Sequencing, Brute Force Engineering, Nanotechnology and Robotics. All these sources have a dual-use potential. They are not only the future's most promising technologies but also are the most dangerous. For example, the gray cloud, a self-replicating airborne nanodevice, theorized by molecular engineer, Ralph Merkle, catalyzes carbon dioxide into graphite. If unleashed, it could result in a solid wall that would cover the earth, block the sun, and eventually destroy life on earth in as little as a few days. As nanotech technology and manufacturing equipment become available for purchase, many more terrorist organizations, groups and countries will be able to make weapons that are currently beyond their reach. Molecular manufacturing and controlling human minds through mind-computer interfaces, although in the formative stages of development, but once they evolve can enable new classes of weapons, for example, (farfetched) tiny flying robots each carrying explosives and that are built with features that can help them to get away undetected.

1.9 Questions for Reflection

A. What role does diplomacy play in controlling WMD?
B. How can the proliferation of WMD to neutral countries be controlled?
C. Which countries own the majority of the world's nuclear weapons?
D. What according to you is the best mechanism of control for avoiding wars of total destruction?
E. How is the production of WMD being controlled?
F. How nanotechnology will affect the role of terrorists in regards to WMD?
G. Why is there a need to rethink arms control?

PART II

Nuclear Weapons—The Pending Peril

2

Nuclear Weapons—The Dangers of a New Era

Nuclear weapons are at the forefront of today's international conflicts. Even a small nuclear device can annihilate the existence of an entire city in no time. Because of their extreme destructive power, nuclear weapons are considered the most deadly WMD.

A nuclear weapon is an explosive device such as a warhead or bomb, which derives its destructive force from the release of nuclear energy—from fission or a combination of fission and fusion. Both these sources release huge quantities of energy from relatively small amounts of matter. A modern thermonuclear weapon weighing little more than a thousand kilograms can result in an explosion akin to the blowing up of more than a billion kilograms of a conventional explosive like TNT (TNT is the short form for trinitrotoluene, which is chemical explosive; TNT is a pale yellow, solid organic nitrogen compound used chiefly as an explosive). One worst-case scenario simulation estimated a one-megaton explosion in Detroit (equivalent to a million tons of TNT) could kill 250,000 people, injure half a million more, and flatten all buildings within a 1.7 mile radius. These are the hardest of all types of weapons to make because the critical weapons-grade nuclear elements of plutonium and uranium are hard to acquire and are very expensive.

Traditional nuclear weapons are not the only threat. The world is concerned that terrorists and rogue states might also target the world's nuclear power plants and supplies.

SPOTLIGHT

The control and use of nuclear weapons has always been a central theme of international policies and treaties since their debut.

In the history of warfare, only two nuclear weapons have been used, both by the United States against the cities of Hiroshima and Nagasaki in Japan near the end of the World War II. The first was detonated on the morning of 6th August 1945, when the United States dropped a uranium gun-type device codenamed "Little Boy" over the Japanese city of Hiroshima. Three days later, on 9th August 1945, the United States detonated a plutonium implosion-type device codenamed "Fat Man" over the Japanese city of Nagasaki. These two bombings resulted in the deaths of approximately 200,000 Japanese people (mostly innocent civilians) from the deadly explosion, searing heat, nuclear fallout and acute radiation sickness. The survivors continue to suffer from long-term effects of (ionizing) radiation even today. These bombings have been widely discussed and are a hot topic in the academic circles ever since. Since these bombings, nuclear weapons have been detonated in over two thousand instances—all for demonstration and testing purposes.

Apart from their use as weapons, nuclear weapons have been used for various non-military purposes, termed as Peaceful Nuclear Explosions (PNEs). For example, Operation Plowshare was the Unites States program for the development of techniques to use nuclear explosives for peaceful purposes. Project Gnome was the first nuclear test of the Plowshare program with the objective of using nuclear explosives for peaceful applications. Germany at one time considered manufacturing nuclear explosives for civil engineering purposes. Nuclear explosions have been studied as a possible method of spacecraft propulsion. According to a latest patented method, atmospheric testing since the 1950s has created new isotopes, Cesium-137 and Strontium-90, which eventually permeated the planet's soil and plant life, and these isotopes can be used to detect art forgeries as natural oils are used as binding agents for paints. The above-ground tests conducted between 1955 and 1980 caused a surge in global levels of radioactive carbon-14, which then rapidly dispersed and equalized around the globe. And it remains in the dental enamel, the hardest substance in the body; radiocarbon dating analysis determines the amount of carbon-14 in the tooth enamel to calculate the age of the tooth and its owner. This technique can be of great significance in forensics science and crime investigation. India's first nuclear device explosion in 1974, "The Smiling Buddha," was described by the Indian Government as a peaceful nuclear explosion.

A WORD OF CAUTION

Nuclear weapons programs, nuclear reactors, uranium mining, background radiation, atmospheric testing and research leave vast amounts of intractable waste, which will be dangerous for the coming generations and the environment, and for which no country has a long-term storage or disposal solution.

2.1 Types of Nuclear Weapons—Fission, Fusion and More

Nuclear weapons can be grouped into different types based on the nuclear reactions and the details of their design. In nuclear weapons, the explosion is created by changing the atoms—they are either split or fused to create new atoms.

There are two basic types of nuclear weapons. The first type derives its explosive energy through nuclear fission reactions alone. Such fission devices are commonly known as atom bombs or atomic bombs—A-bombs. The basic components in an A-bomb include the fuel and some sort of detonation device. A-bombs release energy by splitting the nuclei of uranium or plutonium atoms. In fission weapons, a mass of fissile material, typically enriched plutonium or uranium, is assembled into a super-critical mass—the amount of material needed to start an exponentially growing nuclear chain reaction either by shooting one piece of sub-critical material into another to form one super-critical mass and then a neutron is fired at this mass to start a chain reaction that leads to nuclear explosion, also referred to as the "gun" method, or by compressing a sub-critical sphere of material using chemical explosives to many times its original density also called the "implosion" method. In the implosion method, the weapon is armed with detonators that start the explosion outside of the surface of a regular sub-critical mass, which in turn causes the explosion shock wave to go inward to the fissionable core, and this compresses the core to a super-critical mass that eventually explodes. The former is the simplest technique for assembling a nuclear weapon, whereas the latter method is considered more sophisticated than the former. Both types assemble a super-critical mass of fissile material and use a tamper to hold the core together long enough to produce the desired nuclear explosion.

The second basic type of nuclear weapon produces a large amount of its energy through nuclear fusion reactions. Such weapons are referred to as hydrogen bombs or H-bombs, as they depend on fusion reactions between isotopes of hydrogen, deuterium and tritium. Such fusion reactions require high temperature and pressure levels, which are created by using a fission explosion (i.e., an atom bomb) to trigger the fusion reaction. As extremely high temperatures and pressures are required in order to initiate

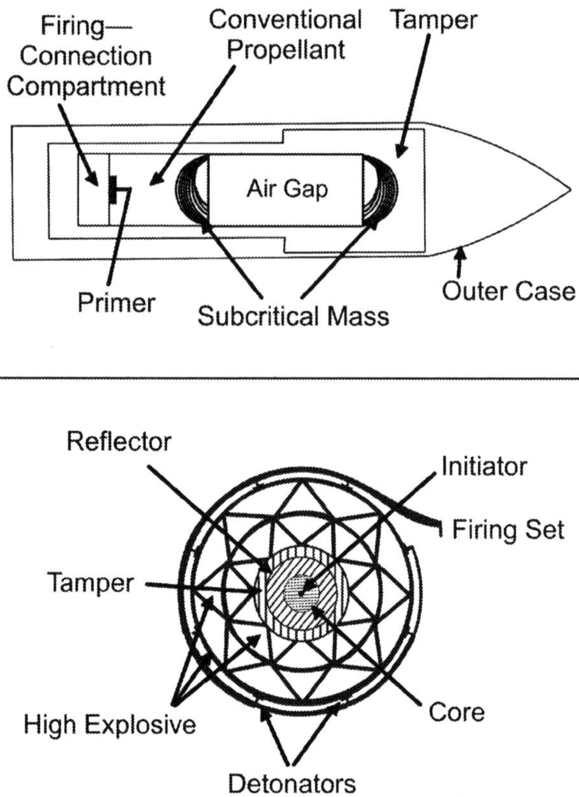

Figure 2.1: The Gun and Implosion Methods

fusion reactions, the H-bomb is also known as a thermonuclear bomb. In fission weapons, uranium or plutonium is split into lighter elements that together weigh less than the original atoms, the remainder of the mass appearing as energy, whereas in hydrogen bombs, joining together of lighter elements into heavier elements occurs. The Teller-Ulam design is the technical concept behind modern thermonuclear weapons; it is the nuclear weapon design concept used in most of the world's nuclear weapons. It contains primarily two stages: Fission bomb (Primary) and Fusion capsule (Secondary). This is done by placing a fission bomb and fusion fuel in proximity within a special, radiation-reflecting container. The detonation starts in the high-explosive layer surrounding the primary. When the fission bomb is detonated, X-rays and Gamma rays emitted first compress the fusion fuel and then heat it to the required thermonuclear temperatures. The resulting fusion reaction produces enormous numbers of high-speed neutrons. These neurons induce fission in the materials and the resulting explosion has a tremendous force with a yield in the order of megatons. In larger hydrogen bombs, the explosion of the fusion secondary can drive yet another larger third stage. Each stage can be 10-100 times the size of the previous stage. The main challenge in the design of these weapons is to ensure that a significant portion of the fuel is consumed before the weapon destroys itself.

A broad level classification of nuclear weapons includes two types—tactical and strategic weapons. A tactical nuclear weapon (or TNW) refers to a nuclear weapon which is designed to be used in a battle against battlefield targets like a naval vessel, facilities and troops, in military operations, etc. TNW includes weapons like artillery, nuclear mines, etc. TNW have used the most variety of delivery types—artillery shells, gravity bombs, torpedoes, nuclear depth charges, land mines and missiles. The modern term for TNW is non-strategic weapon. Their yield and range is generally lower than that of strategic nuclear weapons, and they are used to achieve limited or tactical objectives. For example short-range missiles are tactical, while intercontinental ballistic missiles with nuclear warheads are strategic. As opposed to this, a strategic nuclear weapon (SNW) is one designed for mass destruction like a nuclear missile; it is used against targets like civilian areas and military bases in a strategic nuclear war and according to a strategic plan. SNW is also referred to as high-yield weapons.

Loose Nukes—refers to the nuclear weapons unaccounted for in the former Soviet Union. Before its collapse in 1991, the Soviet Union had more than 27,000 nuclear weapons and enough weapons-grade plutonium and uranium. The collapse of the Soviet Union led to the possibility that former Soviet Union nuclear weapons might become available on the black market and fall into the wrong hands of terrorists and criminals. Today, experts use this term to refer to nuclear weapons, materials or know-how that could fall into the wrong hands. Loose nukes have a spectrum of issues associated with them—nuclear terrorism, nuclear trafficking, nuclear material smuggling deals, proliferation risks, underpaid nuclear scientists selling their know-how, etc. Loose nukes are a worldwide worry even today.

Suitcase bomb or suitcase nuke—it is a very compact nuclear weapon, small enough to fit into a normal-sized suitcase or briefcase. It uses a suitcase as its delivery method. It doesn't take much more than a single critical mass to cause a significant explosion ranging from 10-20 tons. These bombs have been used in terrorist attacks where cell phone-detonated bombs are set off in public places. Its main advantage is that it can pack so much explosive force into such a small space. The lightest nuclear warhead ever acknowledged to have been manufactured by the United States is the W54 and the backpack-carried version called the Mk-54 SADM (Special Atomic Demolition Munition). But as far as the military utility of SADM is concerned, they are not the first choice as military usage requires a high yield, and in this case combining portability with such a high yield is quite difficult.

A boosted fission weapon is a fission bomb that relies primarily on fission, but obtains some of its power from fusion too. In this weapon, a few grams of a deuterium/tritium gas (D-T gas) mixture are included in the center of the fissile core. The fission reaction takes place first and produces the required temperatures and densities for the fusion reaction, and the fusion reaction in turn accelerates the fission reaction. Ignition of D-T fusion generates high energy neutrons which induce more fission in the fission fuel. The boosting process helps to maintain the yields and also decreases the overall size and weight of weapons. "Greenhouse Item" was the first test of a boosted fission weapon conducted on 25th May 1951.

A salted bomb is a nuclear weapon designed to contaminate an area with enhanced quantities of radioactive fallout. Radioactive fallout is the term used for the radioactive particles produced by a nuclear explosion and carried high up into the air by the mushroom cloud, and these particles later on "fall out" of the sky and settle themselves on the ground. The fall out emits long-lasting and lethal gamma radiation thereby rendering the targeted area uninhabitable. When additional elements such as sodium, gold or cobalt are incorporated in a standard weapon, it becomes a salted bomb. The term is from the expression "to salt the earth"—meaning to make an area uninhabitable for generations.

A multiple independently targetable reentry vehicle (MIRV) warhead is a collection of nuclear weapons. It allows a single missile to carry several warheads, each of which can strike a different target. Examples include the Peacekeeper missiles and Trident missiles.

A neutron bomb is a type of thermonuclear weapon. It is technically referred to as an enhanced radiation weapon (ERW). It produces minimal blast and heat, and short-range destructiveness, but a relatively large amount of lethal radiation. It uses fusion to enhance the production of radiation beyond that which is normal for an atomic device. The heavy energy released by the fusion of the neutrons is not absorbed inside but intentionally allowed to escape through an outer shell made of nickel or chromium. Neutron bombs are designed and used for both special and tactical purposes like an anti-missile device, to kill soldiers who are protected by armor, to make armored targets unusable due to radioactive effects, etc. Though this bomb has a lower destructive yield, it is still capable of causing much greater destruction than any conventional bomb. Unlike standard nuclear bombs that cause long-term contamination, the neutron bomb causes less persistent radioactive fallout and the neutron radiation quickly dissipates after the explosion. Samuel T. Cohen of the Lawrence Livermore National Laboratory is the one who developed the concept of this bomb in 1958; he is often credited as the "Father of the Neutron Bomb." Cohen's idea of a neutron bomb was that of it being the ultimate sane weapon. According to him, this bomb would kill people, or as he puts it "the bad guys," but doesn't produce tremendous collateral damage on civilian populations and the environment and infrastructure a civilian population needs to survive. For example, the bomb can be highly effective against troops on the battlefield but would not endanger civilians who are only a few miles away. Although initially opposed by the then American President, John F. Kennedy, its testing was finally authorized, and after a decade of design, its testing was carried out at an underground Nevada test facility in 1963. The first neutron bomb was added to the United States weapons arsenal in 1974. France produced and tested neutron bombs in the early 1980s. Evidence and reports suggest that Russia and China have these bombs too.

2.1.1 *Weapons Design and Production—Materials, Components and More*

Nuclear infrastructure in the form of production facilities, specialized materials, components, weapon system safety controls, and maintenance, inspection and testing equipment like reflector materials, high explosives, high-temperature furnaces, capacitators, fuels, detonators, antennas for receiving radio signals, waste tanks, fuses, process lines, neutron generators, tampers, radiation shields, spin, flow and shear forming machines, detonation devices, firing sets, and photography and recording devices, etc., is needed.

Fissile materials like isotopes of Uranium-235(U-235) and Plutonium-239 (Pu-239) are the most common and the best weapon materials for fission weapons. Fussionable materials include deuterium and tritium.

Advanced weapons design and production require accurate measurements; hence modern tools of computer-assisted fabrication, including computer numerically controlled (CNC) machine tools are needed.

Depending on the type, complexity, sophistication and constraints of the design, the degree of compression needed, timing of the initiation of the chain reactions, etc., additional equipment, know-how and technologies may be required.

2.1.2 *Who has what at a Glance—Current Nuclear Arsenals*

According to estimates by the *Federation of American Scientists*, the United States and Russia are believed to possess the world's largest nuclear arsenal. According to the *World Nuclear Stockpile Report 2011* by Ploughshares Fund (Ploughshares Fund is a foundation that develops initiatives aimed at preventing the use and spread of weapons), Russia is likely to have a total inventory of 11,000 (strategic, non-strategic and operational), whereas the total for the United States is 8,500.

Both India and Pakistan are believed to be increasing their stockpiles. According to *the Bulletin of the Atomic Scientists Report 2011*, the number of nuclear warheads with India is approximately 100 and for Pakistan it is between 90 and 110.

According to the *Arms Control Association (*Arms Control Association is an organization dedicated to promoting public understanding of and support for effective arms control policies*)*, a massive expenditure will be made over the next decade in modernizing delivery systems and warhead production. According to its findings, the nuclear-armed states will spend at least one trillion dollars on nuclear weapons over the next decade.

China's nuclear arsenal could be many times larger than the current estimates. China is well into a very long-term program of growing its arsenal from 240 weapons today to upwards of 250-300 through 2050.

Israel is widely believed to possess nuclear weapons, though it does not declare or acknowledge having them. According to some estimates, Israel possesses as many as 200 nuclear warheads. Some countries such as Brazil, Argentina and South Africa have acknowledged pursuing nuclear programs in the past but have abandoned their programs.

All these numbers are estimates and not absolute official figures because most countries guard their official figures as closely held national security secrets.

Right from small and light warheads for tactical use to new submarines—all point to a new era of growth in the global nuclear arsenals.

2.2 Nuclear Warfare Strategy, Treaties, Agencies—Fight or Avoid

Nuclear warfare strategy is a policy determining the use of nuclear weapons; it is a way for either fighting or avoiding a nuclear war. The policy of nuclear deterrence aims to prevent unwanted action by opponents by convincing them that the resultant costs would exceed any gains. The goal is to always maintain a second strike status—the ability of a nation to respond to a nuclear attack with its own powerful retaliation, and the potential to strive for the first strike status—the ability to completely annihilate an aggressor's nuclear forces before they can retaliate through a preemptive surprise attack. This military strategy gained increased prominence during the Cold War days. During the Cold War, policymakers and military theorists of nuclear-enabled nations developed several policy models with an aim to prevent nuclear attacks on their nations.

The forms of nuclear weapons delivery determine the types of nuclear strategy to implement thereby making it difficult to launch a preemptive strike and defend against them. For example, concealing the location of nuclear weapons, such as burying them in hardened bunkers or hiding the weapons in locations like submarines that are hard to track.

Other responses include a threat of massive retaliation to other forms of deterrence like missile defense to destroy the missiles before they land on the targeted area. There are two types of deterrence strategies—passive or core deterrence and extended or active deterrence. Passive deterrence is the threat with a nuclear-strategic response in case of a nuclear attack on the home territory of the threatening nation, whereas active deterrence threatens with a nuclear-strategic response in case of a nuclear attack on the territory or troops of allies. Of course, in a globalizing age where nuclear weapons are no longer the exclusive property of a handful of powerful states, today the ranks of nuclear powers include small nations like Pakistan too, makes nuclear deterrence less relevant, as there are more non-nuclear options available to strike inter-state reconciliation and compromise. Even in the case of psychotic or suicidal opponents, the various forms of deterrence may not work. Another criticism against deterrence is that it can create a vicious cycle, beginning with political misunderstandings, then mutual distrust and threat, and a subsequent arms race which elevates the risk of actual war. Nuclear arms control must move beyond the deadlock of deterrence to handle current challenges like terrorism.

Table 2.1: Response of Various Countries to Deterrence

Country	*Deterrence Doctrine*
United States	Relations with Russia continue to follow the traditional doctrine of Mutual Nuclear Deterrence. The policy of deterrence towards nations with minor nuclear capabilities should ensure through threats of immense retaliation or even preemptive action that they do not threaten the United States.
Russia	Russia follows the doctrine of limited nuclear war. According to this doctrine, Russia can use nuclear weapons to prevent political pressure against Russia and its allies.
India	India follows nuclear no-first-use doctrine. India maintains that it will not be the first to set off a nuclear strike, but will respond with punitive retaliation should deterrence fail.
Israel	Israel's doctrine is related to its relationships with its Arab neighbors. It includes policies of deterrence to prevent conventional attacks or those involving WMD.

The critics of the idea of nuclear strategy believe that the very idea of waging a nuclear war between two nuclear powers would result in mutual destruction. This viewpoint indicates that the significance of nuclear weapons is purely to deter nuclear attacks because any nuclear attack is triggered out of mutual fear and distrust; hence the result of such an attack would also be mean mutually assured annihilation. This very threat and its magnitude has been a strong motivation for the rise of anti-nuclear weapons activism. Anti-nuclear weapons activism covers a gamut of global grassroots movements working towards the total elimination of nuclear weapons. Unlike conventional weapons, nuclear weapons can manage to successfully deter an all-out war between nations, as they did during the Cold War between the Soviet Union and the United States. But contrary to this view of nuclear weapons, there is yet another standpoint. In the case of suicidal terrorist attacks involving nuclear weapons, mutually assured destruction does not arise and act as a deterrent, as these terrorists are prepared to die for their fanatical beliefs. Again, there is no fixed target or nation to retaliate against, in case terrorists attack first; hence the strategy of deterrence does not work in nuclear terrorism scenarios. Handling such complicated scenarios is the next wave of challenges for nuclear strategies, especially after the September 2001 World Trade Centre and Pentagon attacks.

A decision to use nuclear weapons has several political, military and moral dimensions attached to it. Their immense power and danger has made world leaders and politicians keenly interested in not only how their own nations will use nuclear weapons, but also how other nations are handling their nuclear stockpiles. A ban on nuclear testing is the oldest item on the arms control agenda. Arms control activists have always campaigned for the implementation of a treaty banning all nuclear explosions since the early 1950s. In the late 1940s, the lack of mutual trust and understanding was preventing the United States and the Soviet Union from working out a solution towards international arms control agreements. Later on in the 1960s, measures were taken to control the proliferation of nuclear weapons to other countries and the environmental impact of nuclear testing. This resulted in a string of treaties. The 1963 Partial Test Ban Treaty (PTBT) prohibits nuclear weapons tests or any other nuclear explosion in three environments: in the atmosphere, in outer space and underwater. It does not prohibit underground nuclear explosions. The next major step towards proliferation was the Nuclear Nonproliferation Treaty, which aims to stop the spread of nuclear weapons. Thereafter, not much was done on this front until the end of the Cold War. The next stage in the arms race cessation was a rigorous period of negotiations between parties to the PTBT (1993-1996) with strong support from the United Nations General Assembly to adopt a comprehensive test ban treaty, and the end-result was the Comprehensive Nuclear Test Ban Treaty (CTBT). CTBT is a multilateral treaty that bans all nuclear explosions in all environments, for military or civilian purposes. CTBT seeks to achieve the discontinuance of all test explosions, including underground tests. As of March 2012, 157 countries have ratified the CTBT.

However, this treaty will enter into force only after all the 44 states listed in Annex 2 of the treaty have ratified it.

Table 2.2: Countries and their CTBT Status

Country	*CTBT Status*
The five nuclear-weapons nations under the NPT	
United States	Signatory
Russia	Ratifier
United Kingdom	Ratifier
France	Ratifier
China	Signatory
Non-NPT nuclear powers	
India	Non-signatory
Pakistan	Non-signatory
North Korea	Non-signatory
Undeclared nuclear powers	
Israel	Signatory

The International Atomic Energy Agency (IAEA) is the global agency for cooperation in the nuclear field. It was established as an autonomous organization on 29th July 1957 under the mandate of the United Nations. IAEA works with its member states and multiple partners worldwide to promote safe, secure and peaceful use of nuclear technologies and nuclear energy. It aims to prevent the diversion of nuclear material for weapons purposes through its safeguards system and safeguards agreements. It publishes technical and scientific journals, bulletins and books on nuclear security, international safety standards, nuclear power, radioactive waste management and global nuclear development. It provides a comprehensive hub of websites, databases, publications, safety standards, training material and more. After Japan's Fukushima crisis (a series of equipment failures, nuclear meltdowns and releases of radioactive materials at the Fukushima I Nuclear Power Plant, following the Tôhoku earthquake and tsunami in March 2011), IAEA has planned to strengthen international safety checks on nuclear power plants. In the nuclear Nonproliferation Treaty (NPT) of 1968, the IAEA gained authority for policing the nuclear activities of member countries to ensure that those without nuclear weapons did not acquire them.

2.2.1 *Additional Agreements and Treaties—On-Going Global Efforts*

Additional agreements and treaties have been adopted between individual countries to further the cause of nuclear disarmament and have made it possible to create several Nuclear Weapons-Free Zones (NWFZs) in the world.

The Strategic Arms Limitation Talks Agreement (SALT)—there were two rounds of talks and agreements: SALT I and SALT II between the United States and the Soviet Union. SALT I was signed in 1972 and SALT II in 1979. Both treaties aimed at placing limits and restraints on strategic weapons like long-range or intercontinental ballistic missiles. SALT II was the second treaty between the United States and the Soviet Union resulting from the Strategic Arms Limitation Talks. The goal of SALT II was to replace the Interim Agreement reached in SALT I with a comprehensive treaty on broad limitations on strategic offensive weapons. The United States never ratified SALT II. The treaty faced considerable resistance from the United States Senate. Nonetheless, both the countries followed its terms until 1986. Subsequent discussions took place under the Strategic Arms Reduction Treaty (START) and the Comprehensive Nuclear Test Ban Treaty.

Another important treaty between the United States and the Russian Federation is the New START. It is a nuclear arms reduction treaty, and it seeks to preserve continuity in and provide new momentum to the process of reducing and limiting nuclear arms while maintaining safety of their nuclear arsenals. It also aims to expand this process in the future by adopting a multilateral approach. It limits each side to 1,550 deployed warheads and 700 deployed intercontinental ballistic missiles and submarine-launched ballistic missiles and heavy bombers.

Many countries have been declared as NWFZs. A NWFZ is an area or geographical region where the stationing, production, testing and deployment of nuclear weapons are prohibited. For example, Mongolia and Austria have both declared their nuclear weapons free status through enacting domestic legislation. The Treaty for the Prohibition of Nuclear Weapons in Latin America and the Caribbean (Treaty of Tlatelolco, 1967) prohibits the production or deployment of nuclear weapons in the concerned regions. The South Pacific Nuclear Free Zone Treaty (Treaty of Rarotonga, 1985) places a ban on nuclear explosions and explosive devices for peaceful purposes, and prohibits its members from dumping nuclear waste into the zone's waters.

Other key nuclear arms control treaties are the Southeast Asia Nuclear-Free Zone Treaty (1995), the South Pacific Nuclear-Free Zone Treaty (1986), etc. These are all positive steps towards the on-going global efforts in nuclear disarmament and nonproliferation.

2.3 Delivery Systems—From Traditional to Miniaturization

The systems, military vehicles and technologies used to transport a nuclear weapon to its target are an integral component of both its design and strategy. These weapons are most likely to be delivered as bombs dropped by fly-over bombers or in the form of ballistic missiles. Developing and maintaining delivery systems is one of the most resource-intensive aspects of nuclear weapons. According to a study, since 1940, the United States has spent $5.8 trillion on nuclear weapons programs.

The first method of delivery and the method used in the bombings of Hiroshima and Nagasaki is that of a gravity bomb dropped from a bomber aircraft. This delivery option is the first method developed by countries as the degree of weapon miniaturization is manageable in terms of weapons design expertise and knowledge. It does, however, limit the number of weapons which can be fielded, the response time to an impending attack and the range of an attack. With the advancements in miniaturization, specialized delivery systems are not needed that often as miniaturized versions of high-tech weapons are now very much doable. Miniaturization helps to reduce a component's size and cost to a considerable percentage of their initial values. For example, with miniaturization, nuclear weapons can be delivered by both strategic bombers and tactical-fighter bombers, by allowing an air force to use its current fleet with minimal or no modifications. The majority of the United States nuclear warheads follow the basic weapon and physics package design—the B61 bomb; this weapon can be delivered as a free-fall airburst, a retarded airburst, a free-fall surface burst, or in laydown mode (laydown—delayed action detonation in which the bomb's fall is slowed by a parachute so that it lands on the ground before exploding) from an aircraft. From a strategic point of view, nuclear weapons mounted on launch-ready ballistic missiles are the preferred option. While short-range missiles perform well in regional conflicts in terms of low cost and ease of configuration. Newer delivery options like submarine-launched ballistic missile (SLBM) and intercontinental ballistic missile (ICBM) not only provide the advantage of delivering the missile anywhere on the globe but also have improved accuracy to the point where even the smallest point targets can be successfully attacked. SLBM is a ballistic missile capable of delivering a nuclear warhead that can be launched from submarines. ICBM is a ballistic missile with a long range, capable of traveling from one continent to another. Modern designs usually deliver multiple independently targetable reentry vehicles (MIRVs), allowing a single missile to carry several warheads, each of which can strike a different target. Today, missiles are the most preferred delivery system of nuclear weapons.

The size of a nuclear weapon can be made quite small through miniaturization, which then can easily fit into a truck or large car. This kind of delivery mechanism is a threat in itself as many security experts believe that this delivery method can be easily used by terrorist groups.

2.4 Symptoms and Effects of Exposure—From Damage to Death

Depending on the rate of exposure and the absorbed dosage of radiation, victims may either experience immediate and/or delayed blast effects. The unit used to measure radiation dosage is the rem, which stands for roentgen equivalent in man. Rem helps to determine the biological risks and effects associated with exposure to radiation. The victims can experience anything from vomiting to long-term effects such as cancer, neurological effects, malformed offspring or birth defects. Doses above 100 rems cause vomiting, headache and nausea. Doses of 300 rems or more can cause the loss of white blood cells, hair loss and internal harm. Due to the loss of white blood cells, the victims are vulnerable to hemorrhage and lowered immunity to diseases. Doses above 450 rems are fatal in nature. Many victims do not die directly from the actual explosion but from the radiation released as a result of the explosion, for example, one child who was born two days after the Hiroshima explosion eventually died of acute leukemia at the age of eighteen.

Damage in localized tissue can lead to disturbed organ functioning and cell damage. Higher doses cause gangrene and/or death of localized tissue.

According to Richard Turco, professor at UCLA and expert on the environmental effects of nuclear war, 'For a regional-scale nuclear conflict, fatality estimates range from 2.6 million to 16.7 million.'

Acute radiation syndrome (ARS) occurs within several months of exposure to high amounts of ionizing radiation. Forms of ionizing radiation include: X-rays, Gamma rays, and Alpha and Beta particles. A catastrophic event or accident occurs that results in the exposure of all or most of the body to a high dose of penetrating radiation over a very short period of time, and then causes severe illness. For example, the nuclear reactor meltdowns at Chernobyl and Three Mile Island have spread radiation across Europe and North America, and their effects are felt even today. ARS is also known as radiation sickness and radiation toxicity. The victim goes through various phases following exposure. Without immediate medical attention, survival is highly unlikely.

The blast, heat and fallout are capable of killing almost all people within a two kilometers radius. Again, the number of deaths would depend on factors like the area actually affected, the extent of evacuation and fallout protection, and the location of fallout. For example, in the Hiroshima attack, the vast heat generated by the explosion created a huge upward surge of air, resulting in the mushroom cloud, which contained all the dangerous fission products. These products proved to be a source of grave danger to humans when they returned to earth.

These effects also hamper the ozone layer by depleting it. They also affect the atmosphere and climate for years to come. Based on new work and research by some of the pioneers of nuclear winter research who worked on the original studies, "nuclear winter" will occur even if a minor nuclear war takes place in current times, where large amounts of dust and ash, and soot and smoke (*aerosol* particles) knocked up from the ground into the air can block off sunlight for months, which then would result in sub-freezing temperatures that would lead to a global catastrophe. Nuclear winter is a theoretical concept and the theoretical findings are based on computer model calculations.

The United States National Cancer Institute released a report in 1997 revealing that iodine-131 from nuclear testing was found in every single county of the United States. Another study by Dr. Ernest Sternglass, professor of Radiation Physics, University of Pittsburgh, showed that the Three Mile Island accident led to a minimum of 430 infant deaths.

Major sources of nuclear radioactive material are nuclear power reactors, waste repositories or reprocessing plants. An accident in or an attack on such sources would result in the meltdown and dispersal of the radioactive inventory, and eventually lead to radioactive poisoning.

A nuclear explosion also sends out an electromagnetic pulse, which has the potential to damage or destroy electronic devices over widespread areas, including critical setups like medical equipment.

As there is no known way of neutralizing radioactive materials, apart from sending them through a nuclear reactor, radioactive materials are fatal until they have decayed to such an extent that they no longer release radiation.

From these points, we can conclude that a nuclear explosion and its resultant radiation have prolonged effects on the human health and the ecological balance.

2.5 Responses, Treatment and More

Well-monitored emergency supplies rationing strategies (food, water, first aid and medicines) after the blast need to be instituted in order to ensure survival. People who have inhaled or ingested fallout material need to be administered mobilizing or diluting agents. Medical injuries like burns and illnesses and delayed effects related to radiation exposure like nausea and immune system problems need immediate medical help, psychological support, supportive care and surgical attention.

Shelter is a key response mechanism—storm drains, basements, tunnels and foxholes need to be in place in order to deal with a nuclear blast. Such protection measures are necessary to save people from the fallout, resultant radiation and blast pressure wave. One should stay in these sheltered places at least for 48 hours or as per the instructions issued by the authorities.

SPOTLIGHT

People should move in an inland direction opposite to the site of the blast. Instructions issued by authorities related to medicines, decontamination and evacuation should be strictly followed.

Decontamination is another major response to a nuclear blast. It is necessary to remove radioactive particles and dust from a victim's body, clothing and other objects. Medical trained professionals should carry out proper decontamination procedures in order to ensure proper removal of radioactive particles. Exposure to isotopes that have contaminated the food and water supplies should be avoided and not consumed. Only covered sources of food and water should be consumed.

2.6 Questions for Reflection

A. Why do some countries want to develop nuclear weapons?
B. What happens with the radioactive waste from nuclear power plants?
C. Has any country ever given up its nuclear weapons?
D. Are there any dangers in testing nuclear weapons?
E. How are nuclear weapons different from other WMD?
F. Who was involved in Nuclear Arms Race and what were the causes of it?
G. Has deterrence outlived its usefulness?
H. Why is the whole world worried about the nuclear power of Iran?
I. In which countries are loose nukes a problem?

3

Major Nuclear Powers—Bombs, Warheads and More

The world's current nuclear arsenal is enough to destroy the world several times over. In the subsequent subsections, we will discuss the nuclear weapons of the major nuclear powers like the United States, United Kingdom, Russia and France. Other developing countries like China, India and Pakistan have also managed to leave an indelible mark on the nuclear landscape.

It all began with the United States building and detonating the first nuclear bomb, followed by the USSR. Thereafter, United Kingdom and France followed in their footsteps. The West and the East were not the only factions with nuclear goals and ambitions. Countries like China also wanted to have their nuclear weapons to protect their existence against superpowers. Once China had its own nuclear arsenal, India too developed its own nuclear arsenal, and once India had its own nuclear arsenal, Pakistan too joined the nuclear bandwagon.

Before discussing the nuclear weapons, we need to have a look at their designation system. Prefixes and suffixes are used with numbers to designate nuclear weapons. Weapons in the United States arsenal were designated with Mark ("Mk") numbers until 1968, for example, Mark 4 or Mk4 or Mk-4, and thereafter with "B" numbers, for example B90. "Test Experimental" bombs are designated with "TX". The Rainbow Codes were a series of code names used to disguise the nature of various United Kingdom military research projects. They were mainly used from after the World War II until 1958. After 1958, they were replaced by an alphanumeric code system. Russian weapons are designated RDS, and RDS is associated with several popular full forms like "Russia Does It Alone" or "Reaktivnyi Dvigatel Stalina" (Stalin's Rocket Engine).

Table 3.1: Weapons Designation

Country	*Designation*	*Explanation*
United States	EC	Emergency Capability—a bomb prototype design stockpiled in very limited numbers. For example, EC-16.
	MK, Mk	Mark (MK, Mk)—designation used for bombs until 1968.
	B	B—designation used for bombs after 1968. For example, the MK-57 bomb, after 1968 was re-designated as B57.
	TX	Test Experimental—a prototype bomb or warhead. For example, TX-14.
	W	Missile warheads. For example, the W61 missile warhead.
	XW	Warhead in the development and test phase. For example, XW-41.
United Kingdom	Black	For example, Black Rock—a surface-to-surface guided missile.
	Blue	For example, Blue Streak—a medium-range ballistic missile.
	Brown	For example, Brown Bunny—the codename of a British tactical nuclear weapons project in the 1950s.
	Green	For example, Green Grass—a pure-fission warhead used in Violet Club and Yellow Sun Mk.1
	Indigo	For example, Indigo Hammer—a lightweight plutonium implosion device.
	Jade	For example, Jade River—a continuous wave radar.
	Orange	For example, Orange Herald—a core-boosted large fission weapon.
	Pink	For example, Pink Hawk—an air-to-air missile.
	Purple	For example, Purple Granite—a two-stage H-bomb.
	Red	For example, Red Beard—the first tactical nuclear weapon.
	Violet	For example, Violet Club—the first operational bomb and an interim megaton weapon.
	Yellow	For example, Yellow Sun—the first operational high-yield strategic nuclear weapon.
Russia	RDS	For example, RDS-1—the first Russian *atomic* bomb and was used in Russia's first nuclear weapon test, codenamed Joe-1.

3.1 United States—The Pioneer in Nuclear Weapons

The United States is the first country to develop nuclear weapons. The United States is the only country to have used nuclear weapons against another nation in an actual war—bombings against Japanese cities of

Nagasaki and Hiroshima in 1945. This is when the nuclear weapons age and the nuclear arms race began.

The earliest bombs were heavy and could only be carried by large bombers. By the early 1950s, the United Sates had managed to successfully design lighter-weight weapons that could accommodate any role and mission, right from short-to-intercontinental-range attacks to anti-submarine warfare missions.

At the end of the Cold War in 1991, the United States had an active arsenal of 23,000 weapons of 26 major types. Since that time the actual nuclear warhead production has been ceased in the United States, although retrofit, modification and maintenance is still going on. According to the Energy Department plans, 4,300+ warheads are scheduled to be retired for dismantlement by 2012. According to latest news, proposed further cuts are in the offing.

The United States and Russia have always kept a close eye on each other's nuclear arsenals and this tab is the reason for international discussion and the introduction of various treaties between the two nations (for example, SORT agreement, START treaty). Despite all these agreements and treaties, both nations have a big *post*-Cold War stockpile.

The size of the Unites States arsenal in terms of numbers and the destructive power has always been the subject of international debate for many years. As the United States has always been a forerunner in the nuclear arms race, its long-term policies and intentions have always played a deciding role in determining the stands taken by other nuclear powers and the future landscape of nuclear weapons.

3.1.1 *Little Boy—The First Nuclear Weapon Used in Wartime*

The **Little Boy**, also known as Mk-1, was a gun-triggered fission bomb that was dropped on Hiroshima by the American B-29 bomber, the *Enola Gay*, piloted by Colonel Paul Tibbets, on 6th August 1945. It was the first nuclear weapon used in wartime. Dimensions: 10 feet long, 28 inches in diameter and a weight of 9,700 pounds. It had a 14.5 kiloton yield (estimates vary; different sources quote different yields). The exact specifications remain classified and are a top secret even today as they could still be used to create more weapons. But still, based on speculations, photographic evidence and interviews with former Manhattan Project personnel, speculated specifications have been derived.

Little Boy was a product of a covert operation known as the Manhattan Project during the World War II. No test of a gun-triggered device had occurred before the Little Boy explosion on Hiroshima.

Although it was used occasionally in experimental devices, the gun method was only used once as a weapon because of the fact that this bomb's design was unsafe and did not incorporate safety measures to handle accident scenarios like accidental detonation. It incorporated only basic-level safety measures. With this design flaw, there were chances of it going off accidentally in any one of the following scenarios—electrical short circuit, crash, fire, lightning or misfire.

Design and Assembly—the bomb derived its explosive power from the nuclear fission of uranium 235. The fissile material, U-235, was divided into two parts, the 'projectile' and the 'target.' The bomb was designed based on the gun-type method of shooting a piece of sub-critical U-235 into another piece of U-235 to create a super-critical mass. Once the two pieces of uranium were brought together, the initiator would start a burst of neutrons and the chain reaction would begin, continuing until the energy released became so great that the bomb simply would blow itself apart. It was the first uranium-based detonation. Unlike the "Fat Man" bomb, the Little Boy design was never tested at the Trinity test site, as enriched uranium was very rare at the time and the United States wanted to conserve its uranium inventory. Moreover, in comparison to the implosion method, the gun-type assembly was considered to be much simpler and considered almost certain to perform. Its design relied on shooting one piece of U-235 (the uranium projectile would be fired through a gun barrel) into another piece by means of chemical explosives to create a nuclear chain reaction.

Construction and Delivery—the heavy cruiser USS Indianapolis transported the parts and enriched uranium fuel to the Pacific Island of Tinian on 14th July 1945. While returning from this mission, Indianapolis was sunk by a Japanese submarine. On 26th July, the three sections of the uranium target assembly were shipped from Kirkland Air Force Base near Albuquerque, New Mexico. Thereafter, with all the necessary components delivered to Tinian, the final weapon was assembled and readied by 1st August. On 6th August, the bomb was loaded into the Enola Gay, a B-29 piloted by Col. Paul Tibbets. Enola Gay made the 750-mile trip to Japan and dropped the bomb on Hiroshima.

Blast and Effects—the Hiroshima blast is seen as a decision that the United States made during the World War II in order to end the war with Japan. The bombing was spared conventional bombing with a purpose of studying the effects of a nuclear bombing on an undamaged city. The bomb literally destroyed everything in its path; the explosion wiped out 90 percent of the city. The bomb was dropped by a parachute at 8:15 in the morning and the detonation occurred at an altitude of approximately 1,980 ft. As per the published figures from the United States Army, 66,000 people were killed as a direct result of the blast and 69,000 were injured. The following effects were experienced: blast, primary (caused by the heat radiated from the atomic explosion) and secondary fires (for example, fires resulting from the damage to electrical systems), flash burns from primary heat waves, and radiation and mechanical injuries. Deaths from radiation were very common for years after the explosion. This blast brought about effects that have crippled Japan for many years. The center of the city was hit and everything but the reinforced concrete buildings had virtually disappeared.

3.1.2 *Thin Man—A Proposed Experimental Gun-Type Plutonium Bomb*

The **Thin Man**, formally known as Mark 2 nuclear bomb, was a product of the Manhattan Project. It was a proposed experimental gun-type plutonium bomb. Estimated dimensions: 2 feet wide and 18 feet long with a weight of 7,500 pounds.

Initially, the Manhattan Project team believed that 2 pieces of sub-critical plutonium could simply be shot into each other to create the explosion similar to the nuclear reaction created in the gun-type Little Boy bomb, and this is when the idea of the Thin Man came into being The intention of this weapon was to provide a gun that would act as a delivery system for a nuclear bomb; a gun that could fire a plutonium bullet across great distances at more than 3,000 feet per second.

Despite rigorous efforts and modifications from the scientists, this weapon was canceled because test firing of prototypes found the gun to be unstable and unreliable as the plutonium being used was not pure enough to create the desired results and it had a high spontaneous fission rate that could lead to pre-detonation. Moreover, the risk of exposure to radiation was high. It was eventually canceled in 1944 in favor of the idea of an implosion-type bomb.

3.1.3 *Fat Man—The Second Nuclear Weapon Used in Wartime*

The Hiroshima bombing failed to elicit immediate Japanese surrender, therefore, three days after the Hiroshima bombing, on 9th August 1945, Major Charles Sweeney flew another B-29 bomber, *Bockscar*, carrying the "Fat Man," also known as the Mark 3. The 10,200 pounds weapon was dropped over the city of Nagasaki at 11:02 AM from an altitude of about 1,650 feet. Dimensions: 60 inches in diameter, 128 inches long and had a yield of 21 kilotons TNT.

The gun-type design proved to be suitable for uranium-based bombs and not for plutonium. As a result, the Los Alamos Laboratory began developing an implosion design for a plutonium-based weapon and the end-product was the Fat Man weapon. It was an implosion-type weapon with a plutonium core (plutonium-239) similar to the experimental device "the Gadget."

Design and Assembly—the Fat Man was a more complex weapon than Little Boy, thus it needed more testing to gain experimental data. It was the first implosion-type atomic weapon. It used the implosion method, where a ring of detonators shoot segments of plutonium together to obtain the super-critical mass, which then eventually creates a nuclear explosion. Its element base was plutonium. A sub-critical sphere of plutonium was placed in the center of a hollow sphere of high explosive. Several detonator pairs (a ring of 64 detonators) located on the surface of the high explosive were fired simultaneously to produce a powerful inward pressure on the core. This pressure squeezed the core and increased its density, which then resulted in a super-critical state and explosion.

Construction and Delivery—the plutonium core and its initiator were flown from the Kirkland Air Force Base to the Tinian Island in a C-54 cargo plane on 26th July 1945, and these components arrived on the 28th July. On 28th July, three B-29s delivered three bomb assemblies, including units F-31 and F-32. These components reached the Tinian Island on 2nd August. On 8th August, the assembly of unit F-31 with the plutonium core was completed. The original bombing date was set to 11th August and the primary target was Kokura, but the forecast of bad weather around this date moved the bombing date up to 9th August. The primary target was the city of Kokura, but on day of the attack, obscuring clouds and a smoke cover resulted in changing the course to the alternative location of Nagasaki, the only secondary target in range.

Blast and Effects—the number of victims and the damage at Hiroshima was much higher as Hiroshima was on flat terrain while the topography of Nagasaki was hilly in nature. Due to the hilly terrain of Nagasaki, the damage caused by the detonation was less extensive in comparison to Hiroshima. 39,000 people were killed as a direct result of the blast, 25,000 were injured, and several thousand suffered from residual material from the weapon and radiation poisoning. The bombing destroyed 39 percent of all the buildings. 70 percent of the city's industrial zone was devastated. The Mitsubishi Arms Manufacturing Plant, the major military target in Nagasaki, was completely destroyed. The various sources and published estimates put fatalities from 22,000 to 74,000. The main effect was the surrender of the Empire of Japan. On 2nd September 1945, the Japanese government surrendered unconditionally.

3.1.4 *Mark 4—The First Assembly-Line Produced Nuclear Weapon*

True mass production of nuclear weapons began with the **Mark 4** (also referred to as Mark-IV, Mk-4 and Mk 4) in 1950. After the end of the World War II, work at Los Alamos National Laboratory (previously known as Los Alamos Scientific Laboratory) focused on developing a completely new implosion system, which resulted in the design of the Mark 4. It was the first weapon to be produced on an assembly line technique and to be stockpiled in large numbers. The Mark 4 nuclear bomb was an American nuclear bomb design produced in 1949 and these bombs were finally taken out of service in 1953. A total of 550 Mark 4 weapons were produced. It was replaced by the Mark 6 bomb.

This weapon was a redesign of the earlier Mark 3 Fat Man, used in the Trinity test and Nagasaki bombing. The previous design was essentially designed as an emergency wartime expedient weapon whereas the Mark 4 was an enhanced and reengineered version of the basic design in terms of safety, ease of stockpiling, ease of manufacturing, handling and reliability.

The Mark 4 had a diameter of 60 inches and a length of 128 inches. The versions weighed 10,800 to 10,900 pounds. The various models and pit options gave nuclear yields of 1, 3.5, 8, 14, 21, 22 and 31 kilotons. It weighed slightly more than Mark 3. This weapon used composite uranium and plutonium fissile material core or pit; it used type C and D pits.

It introduced the concept of IFI, short form for in-flight insertion, wherein with the help of a special detachable device, the nuclear core of active material could be inserted without complete disassembly. Several enhancements were made to streamline its functioning. The fusing and firing systems were improved, the shape of the case was changed in order to achieve better ballistic accuracy and the outer casing had been streamlined to avoid the aerodynamic wobble.

The **W4** missile warhead, a Mark 4 bomb derivative, was the planned warhead for the Snark SSM cruise missile. It was designed but never built; this design was canceled in 1951.

3.1.5 Mark 5—High Efficiency Implosion Bomb

The **Mark 5** nuclear bomb and the **W5** nuclear warhead were a common core nuclear weapon design designed in the early 1950s. They were pure fission weapons.

The Mark 5 was in service from 1952 to 1963 and the W5 was in service from 1954 to 1963.

The Mark 5 bomb was the first attempt to design lightweight nuclear weapons for tactical purposes. This design was in service from 1952 to 1963. It used a composite uranium and plutonium fissile material core or pit. It was designed to be armed with automated In-Flight Insertion; it was the first weapon to use auto IFI. It was the first weapon with major size and weight reductions over the Fat Man.

The **W5** was the warhead for the Matador (MGM-1) and Regulus 1 (SSM-N-8) SSM cruise missiles. The versions of the Mark 5 bomb weighed

3,025 to 3,175 pounds. The versions of the W5 warhead weighed 2,405 to 2,650 pounds.

3.1.6 Mark 6—Improved High-Yield, Lightweight Weapon

The **Mark 6** weapon was the first atomic bomb to be mass produced by the United States. It looked almost identical to the Mark 4 and Mark 3 weapon designs, but had enhancements. It was produced from 1951-1955; it underwent seven modifications Mod 0 to Mod 6. These bombs were finally taken out of service in 1962. They were removed from the nuclear inventory and their nuclear material was recycled into the more powerful H-bombs, as these required only a minimum amount of nuclear material to produce massive explosion in comparison to the Mark 6 which required a great deal of very expensive nuclear material to produce a massive explosion.

It could be configured for various yields—8, 26, 80, 154 and 160 kilotons. It weighed less than its predecessors as the bomb casing was made of aluminum instead of steel. This bomb was far more aerodynamically stable; hence more accurate. It was also equipped with improved ballistic characteristics. It was 128 inches long and 61 inches in diameter, and weighed between 7600 and 8500 pounds depending on the core used. It could be delivered by the B-29, B-50, B-36, B-47 and B-52 Armed Forces aircraft, and also by the Navy aircraft like AJ-1.

This bomb was a capsule bomb, wherein the nuclear material for the bomb was kept in a special capsule separate from the rest of the device for safety purposes. Just before the bomb was to be dropped from the delivery aircraft, the weapon would became armed by inserting the capsule into the bomb casing.

Variants—Mark 6 served as the basis for the development of the Mark 13 and the Mark 18 nuclear bombs.

In 1956, during a training session, a B-47 bomber crashed into a nuclear weapons storage facility housing three Mark 6 bombs at the Lakenheath Air Base in Suffolk, England. These weapons were quickly engulfed in flames by burning jet fuel, but the fire fighters were able to extinguish the jet fuel before it ignited the TNT.

3.1.7 Mark 7—The First Multipurpose Tactical Weapon Used by all Three Forces

The **Mark 7 'Thor'** was the first multipurpose tactical bomb and was the first weapon used by all three armed services. It saw service from 1952 to 1968. It was tested in the Operation Buster-Jangle test in 1951; this was a test of the TX-7E, a Mark 7 bomb prototype.

It used the 92 lens implosion system. Its element base was made of composite uranium and plutonium. It was capable of either an air or ground detonation. It used a new delivery method, the Low Altitude Bombing System (LABS). A method of bombing where an aircraft flies on a line towards the target and at a predetermined point pulls up sharply. The LABS automatically releases the weapon at the proper angle and speed, and lobs it towards the target. The bomb then would follow a ballistic trajectory to the target allowing the carrier aircraft time to leave the blast area safely.

It had five yields—8, 19, 22, 30 and 61 kilotons. It weighed 1680 pounds. It had 10 different Mods.

Initially, only aircrafts such as the B-29 Superfortress, the all-jet B-47 Stratojet and the B-36 Peacemaker were capable of carrying atomic bombs that weighed upwards of 10,000 pounds, but with the lightweight, compact-sized Mark 7, nuclear capability was added to smaller, faster aircrafts as well. Naval aircrafts like the A-1, A-3, F2H, F3H-2N and FJ-4B carried the Mark 7, and Air force aircrafts like the B-45, B-57, F-84, F-100 and F-101 also delivered this weapon.

The **W7** warhead—the multipurpose warhead version was used in the Navy's Bureau of Ordnance Aircraft Rocket (stockpiled from 1956 to 1963) and the *Mark 90* (alias *Betty)* depth charge (stockpiled from 1955-1950), and in the Army's MGM-5 Corporal (stockpiled from 1955 to 1960) and MGR-1 Corporal Honest John missiles (stockpiled from 1954 to 1960). The manufacturing of the W7 began in 1953 and it retired in 1967. It weighed at 900 to 1,100 pounds depending on its versions.

3.1.8 Mark 8—Early Earth-Penetrating Weapon

The United States fielded two early earth-penetrating weapons in the 1950s, the Mark 8 and Mark 11 bombs. The **Mark 8** bomb was designed in the late 1940s and the early 1950s. It was codenamed as "Elsie" for LC or light case. It was a gun-type highly enriched uranium (HEU) weapon. It was the first gun-assembly HEU bomb produced since the "Little Boy." It had a yield of 25 to 30 kilotons. It could be could be carried internally or externally by a variety of carriers: AD, AJ, A3D, F2H, F3H and F7U-3. It was in service from 1952 to 1957. It was replaced by the improved Mark 11 bomb.

Mark 8 was the first weapon designed to detonate after hitting the ground. It was intended for digging into the ground some distance before detonating. It was intended for destroying hardened targets like enemy submarines in sheltered pens, underground command posts and armored ship decks. This weapon could reportedly penetrate hard sand, reinforced concrete, clay and armor plate.

It was considered as a cratering warhead for the *SSM-N-8 Regulus* cruise missile. In 1951, authorization was given by the Atomic Energy Commission to develop a variant, the **XW-8/Regulus** warhead, but it was canceled in 1955.

3.1.9 Mark 10—Lighter Mark 8 Variant

The **Mark 10** bomb was a lighter Mark 8 variant, both in terms of size and weight. It was codenamed "Airburst Elsie." Just like its predecessor, the Mark 8, this weapon too was a gun-type nuclear device.

It had a yield of 12 to 15 kilotons, and weighed 1,500 or 1,750 pounds. It was intended to be a general purpose airburst (surface target) nuclear weapon.

This bomb was canceled in 1952 and was superseded by the Mark 12 nuclear bomb implosion design, which used considerably less fissile nuclear material and was much lighter.

3.1.10 Mark 11—Improved Version of the Mark 8

The **Mark 11**, an earth-penetrating weapon, was an improved but slightly heavier version of the Mark 8. It was a safety upgrade that replaced the Mark 8 in May 1957. The production of Mark 11 began in 1956; its production was limited to replacing the Mark 8 weapon on a one-for-one basis. It was stockpiled as the *Mk-91 penetration bomb* and it saw service till 1960.

Just like the Mark 8, it was a gun-type bomb. The Mark 11 had a much more advanced external casing designed to penetrate further and more reliably into the ground. It weighed at 3,210 to 3,500 pounds. It had a yield of 25 to 30 kilotons.

It had a fuzing option of Pyrotechnic delay in which there was a time delay between the actual firing and detonation of the weapon.

3.1.11 Mark 12—A Major Step Forward in Miniaturization

The **Mark 12** weapon was a slender, lightweight tactical nuclear weapon. It was an improved version of the Mark 7. It was nicknamed "Brok." It entered the national inventory in 1954 and had an operational run until 1962. It was produced in 3 Mods.

This bomb was one of the first really superior designs. It was a major step forward in weapon miniaturization and the lightweight factor with an implosion system diameter of 22 inches weighing only 550 pounds and a total weapon weight of 625 pounds. It had a yield of 12 to 14 kilotons. It could be delivered by the AJ, A3D, A4D, F9F-8B, F3H, F7U-3 and FJ4-B.

It was the first lightweight bomb that could be carried and deployed at supersonic speeds. It was a much more fissile material efficient design than its predecessors like the Mark 10. It had a yield of 9 kilotons. It was probably the first nuclear weapon to use beryllium as a reflector-tamper inside the implosion assembly. It was equipped with advanced features like levitated fission pits, 92-point detonation and more explosive lenses.

There was a planned *W12 variant,* which would have been a nuclear warhead for the long-range surface-to-air missile, RIM-8 Talos, but it was canceled prior to its introduction into operational service.

3.1.12 Mark 13—High-Yield Mark 6 Follow-On Weapon

The **Mark 13** was a high-yield Mark 6 follow-on weapon. It was roughly the same size as that of the Mark 6. The Mark 13 nuclear bomb and the **W13** missile warhead were developed as improved versions of the Mark 6. They were of the same size and basic configuration as the Mark 6, but utilized an improved 92-point implosion system. Both the Mark 13 weapon and W13 warhead were canceled.

The bomb was 128 inches long and 61 inches in diameter. It weighed 7,400 pounds. The W13 warhead was slightly lighter; it weighed 6,000 to 6,500 pounds. The bomb version and the warhead version were experimental nuclear weapons.

The successful developments in thermonuclear weapon design made this weapon obsolete; it was never deployed and was canceled in 1953. The Mark 13 bomb was superseded by the Mark 15 and the Mark 39 bombs.

The warhead was intended for the Snark cruise missile and the Redstone intercontinental ballistic missile, but was eventually canceled in 1953 and superseded by the W15 and W39 warheads.

3.1.13 Mark 14—The First Solid-Fuel Staged Hydrogen Bomb

The **Mark 14** bomb was the first solid-fuel staged hydrogen bomb. It was also called the EC-14 ("Emergency Capability") or TX-14 ("Experimental") or the "Alarm Clock" device (unrelated to the Alarm Clock/*Sloika* weapon design).

It was a part of the first generation of the United States thermonuclear weapons (along with the Mark 16 and Mark 17). It was one of the first emergency capability thermonuclear weapons. It was an experimental design and as it needed 95% enriched Lithium isotope 6 lithium deuteride, which was costly and scarce at the time, only a limited number (approximately five units) were produced and stockpiled in 1954. This weapon design was proof tested during the Operation Castle Union shot of 16th April 1954. RACER IV was the version used as the primary in the test.

It could be carried by the B-36 and B-47 bomber. The design yield was 6.9 megatons. This weapon weighed between 28,950 and 31,000 pounds. It had a length of 222 inches and a diameter of 61.4 inches.

This weapon was soon retired due to the cost factor. By 1956, the components of all the Mark 14 weapons had been recycled into the more versatile Mark 17 bombs.

In 1998, while remediating a classified landfill, the Sandia National Laboratory recovered a casing of this bomb. It was sent to the classified DTRA Weapons Display Area for refurbishment.

3.1.14 Mark 15—The First Relatively Lightweight Thermonuclear Bomb

The **Mark 15** weapon was the first relatively lightweight thermonuclear bomb. It weighed 7,600 pounds. It was in service from 1955 to 1965. It was widely known as a transitional design between fission and thermonuclear weapons. Unlike most modern thermonuclear bombs, the Mark 15 used a secondary which was primarily highly enriched uranium (HEU). In this bomb, most of the energy came from the HEU fission process once the primary imploded it.

There were three variants: Mod 1, Mod 2 and Mod 3. All the variants had similar specifications in terms of size and weight, but they did vary in their yields.

The **W15** warhead was a Class "C" missile warhead derived from the Mark 15 and intended for use on the SM-62 Snark missile. It was canceled in 1957 in favor of the W39 warhead.

In an accident during a military training exercise session in 1958, this bomb was released after the B-47 carrying it was severely damaged in a mid-air collision with an Air Force F-86 Saberjet off the coast of Savannah, Georgia. This bomb was dropped off the Coast of Tybee Island to reduce weight and to prevent it from exploding during an emergency landing.

3.1.15 Mark 16—Weaponized Version of the Ivy Mike Device

The **Mark 16** was a thermonuclear bomb and it was a weaponized version of the "Sausage" design tested in the Ivy Mike shot. It was also known as EC-16 ("Emergency Capability") or TX-16 ("Experimental"); it only existed in Emergency Capability/Experimental versions. It was the only cryogenic thermonuclear weapon ever deployed; it used a cryogenic liquid deuterium fusion fuel.

This weapon was a joint effort of the TX-16 panel, which consisted of experts from the Air Force Special Weapons Center, Sandia Laboratory and Los Alamos National Laboratory. Each team was assigned with specific tasks—the Air Force Special Weapons Center was responsible for providing a suitable carrier and handling equipment, the Sandia Laboratory was responsible for the parachute, contact fuze, power supply and test devices, and the Los Alamos National Laboratory was responsible for overseeing the overall project.

It had a yield of 6 to 8 megatons. It had a diameter of 61.4 inches and a length of 296.7 inches. It could be carried by the B-36 and the B-47 aircraft.

A unit of the TX-16, code named "Jughead," was scheduled to be tested during the Castle series in 1954, but it was deleted from the test schedule and this test was canceled. The successful test of the solid-fueled *Shrimp* device in the Castle Bravo test resulted in the cancelation of the Jughead test and eventually the cancelation of the EC-16 program.

Only a limited number was produced—5 units were produced. They entered the inventory in January 1954 and were taken out of service by April 1954. Such a short service period was due to the fact that this weapon required cooling systems and cryogenic temperatures, and moreover, the performance of solid fuel weapons proved to be highly effective. This weapon was replaced by the Mark 17.

3.1.16 Mark 17 and Mark 24—The First Mass-Produced Hydrogen Devices Deployed

The **Mark 17** was manufactured between 1954 and 1955. It had a short service life (1954-57); these bombs were withdrawn from service by August 1957. The reason for such a short service life was believed to be its immense size, which was also confirmed by pilots of the B-36 (the carrier for Mark 17). Once the B-36 was removed from the inventory, so was the bomb. This bomb had an explosive force in the megaton range. At 11 megatons the Romeo shot of 26th March 1954 was the third largest test ever detonated by the United States. The Runt I device in this shot was a proof test of the Mark 17 bomb. In 1957, this bomb was involved in an accidental loss incident. This bomb fell out of a B-36 bomber, which was ferrying the weapon from the Biggs Air Force Base, Texas to the Kirtland Air Force Base, Albuquerque. Although the weapon's parachute was deployed, it failed to fully retard the weapon's fall. Due to the low altitude level, the weapon detonated on impact. This accident created a crater 12 feet deep and 25 feet across, and it also spread radioactive fuel material and debris over a wide area. The unfortunate event was characterized by the Department of Defense as a "Broken Arrow" incident.

The **Mark 24** was a bomb design based on the third American thermonuclear bomb test, Castle Yankee. 105 were produced between July 1954 and November 1955. It was retired in 1956. It had a yield of 10 to 15 megatons.

The Mark 17 and Mark 24 were the first mass-produced hydrogen devices deployed by the United States. They were by far the largest air droppable nuclear devices operated by the United States. Both these bombs were based on the capsule design, where the bomb's fission primary was inserted only prior to the weapon usage. The basic versions of the Mark 17 and the Mark 24 were deployed as part of the "Emergency Capability (EC)" program; they had no safety measures like In-Flight Insertion or a parachute retarder. These EC bombs were then replaced with the Mark 17 Mod 0 and Mark 24 Mod 0 bombs in 1954. These advanced versions were equipped with a 64-foot parachute used to stabilize the bomb when dropped and slow down its descent in order to give the bomber greater time to escape from the target area safely. These were then upgraded to Mod1 with the In-Flight Insertion safety mechanism. Some of them were upgraded to the Mod 2 standard when they were equipped with a contact fuze.

The Mark 17 and the Mark 24 bombs were identical to each other in appearance. The only difference between the two was in the bomb's primary. The bombs were 24 feet 8 inches long and 61.4 inches in diameter. They weighed 21 tons. Both these bombs were withdrawn from service in favor of smaller and lighter weapons such as the Mark 15.

3.1.17 Mark 18—The Largest Pure Fission Weapon Design

The **Mark 18**, also known as "SOB" (Super Oralloy Bomb), was a very high-yield fission weapon. It was the largest pure fission weapon design ever developed and tested by the United States. It had a yield of nearly 500 kilotons. It was successfully tested at its full yield in the Ivy King test at the Enewetak Atoll in the Pacific Ocean in 1952. Ivy King was a prototype of the Mark 18 weapon. It was in service from 1953 to 1956.

This weapon was based on its predecessors, the Mark 3, Mark 4, Mark 6 and Mark 13. It was a re-engineered version of the Mark 6 with improved safety measures and a 92-point implosion system. As the bomb had over four critical masses of fissile material in the pit, it was essential to have a safety mechanism in place. For preventing accidental or premature explosion, an aluminium/boron chain was placed in the center, which would not only absorb neutrons but also control the process of arming the weapon.

Mark 18 had a unique fissile pit containing over 60 kg of pure highly enriched uranium as opposed to the normal mixed uranium-plutonium pit. This device weighed 8,600 pounds.

It had a short operational run from 1953 to 1956. All Mark 18 bombs were modified into the lower yield Mark 6 bomb variants (Mark 6 Mod 6) in 1956.

3.1.18 Mark 20—Planned Successor to the Mark 13

The **Mark 20** was a planned successor to the Mark 13; it was intended to be an improved high-yield Mark 13. It was the same size as the Mark 13, but it weighed only 6,400 pounds. Its development was canceled at the same time as the Mark 13.

3.1.19 Mark 21—Weaponized Version of the Shrimp Design

The **Mark 21** bomb was a gravity bomb developed and deployed immediately after the Castle Bravo test in 1955. It was a redesigned Shrimp device with 95% enriched Lithium-6 fuel. It was a weaponized version of the TX-21 "Shrimp" prototype, the first "dry" (lithium deuteride) fuel thermonuclear bomb detonated in the Castle Bravo thermonuclear test in 1954. The weaponization process commenced three weeks after the test and design stabilization was achieved in mid-July of 1955 with a projected yield of 4 megatons, which was subsequently proof-tested as "Clean" Mk-21C at 4.5 megatons in Redwing Navajo in 1956.

This weapon weighed about 15,000 pounds; it weighed less than half the weight of Mark 17 and Mark 24 bombs. It was significantly smaller than its predecessors. During June-November 1957, all older Mark 21 bombs were converted to Mark 36 Y1 Mod 1 bombs.

The **W21** warhead variant was a potential warhead for the SM-64 Navaho missile and the weapons pod of the B-58 bomber. It would have used the physics package of the TX-21 bomb, but this warhead design was canceled in 1957 and was replaced by the much smaller and lighter XW-39. Though the bomb version was stockpiled for a brief time period, no warheads were ever constructed.

3.1.20 TX-22—UCRL Design Based on the Morgenstern Device

The **Mark 22** or the **TX-22** bomb was a University of California Radiation Lab (URCL which is now Lawrence Livermore National Laboratory) design based on the Morgenstern/Ramrod devices. It was canceled in 1954 due to the failure of the "Morgenstern" device which was tested in the "Koon" shot of Operation Castle. Following the Morgenstern fizzle because of premature heating of the secondary from exposure to neutrons, the TX-22 was terminated.

3.1.21 Mark 26—A Mark 21 Sibling Design

The **Mark 26** bomb was related to the Mark 21; it was a Mark 21 sibling design. Its development began in 1954, but was terminated in 1957. This cancelation coincided with the decision of not to pursue the clean version of the Mark 21. It is believed that the Mark 26 was the designation for the clean version of the Mark 21.

3.1.22 Mark 27—Largest-Yield Free Fall Bomb Deployed by the Navy

The **Mark 27** was the largest-yield free fall nuclear bomb deployed by the navy. It was a University of California Radiation Lab (URCL which is now Lawrence Livermore National Laboratory) design. This design was a competitor with the Mk-28 to meet the Class "D" lightweight thermonuclear bomb requirement. The bomb had three versions. It weighed 3,150 to 3,300 pounds. It could be delivered using the Low Altitude Bombing System (LABS). It was designed to be carried by A-3 Skywarrior and A-5 Vigilante bombers. It had a yield of 2 megatons.

The **W27** warhead was the warhead derivative of the first lightweight thermonuclear bomb. It had a yield of several megatons and weighed 2,800 pounds. It was used for the SSM-N-8 Regulus cruise missile. It was also considered for several other platforms like the Regulus II, Rascal and Matador cruise missiles, and the B-58 and F-101 bomb pods, but all these systems were canceled.

The Mark 27 bomb and the W27 warhead were produced from 1958. Both were retired from service by 1965.

3.1.23 Mark 28—Longest Service Duration

The **Mark 28** hydrogen bomb was first produced in 1958. It was in service from 1958 until it retired in 1991. After 1968, it was redesignated **B28**. It had the longest service duration of 33 years.

It was produced from 1958 through 1966. Its production began in 1958, but there was a delay of 5 months due to a glitch in its safety mechanism. After corrections were made, production continued.

It was capable of a ground or air burst and could be carried externally or internally, with a free-fall or parachute retarded drop. It was a multipurpose tactical and strategic bomb carried by virtually all Strategic Air Command bombers. It had 4 Mods and variable yields depending on the versions—Mod 1 had a yield of 1.1 megatons, Mod 2 had a yield of 350 kilotons, Mod 3 had a yield of 70 kilotons and Mod 5 had a yield of 1.45 megatons.

It used the "Building Block" principle, which allowed various combinations of components in terms of size, fuzing and delivery options, configurations and yields. The main combinations were:

A. **B28IN (IN stands for INternal):** Carried internally by B-52 and F-105; unstreamlined internal-carriage version for free-fall delivery; no parachute.
B. **B28FI (FI stands for Full-Fuzing Internal):** Carried internally; unstreamlined internal-carriage version with parachute for laydown

delivery; equipped with one 4-foot pilot chute, one 16-foot and one 24-foot chute.

C. **B28EX (EX stand for EXternal):** Carried externally by F-100, F-105 and F-4; streamlined external-carriage version for free-fall delivery; no parachute.

D. **B28RI(RI stands for Retarded Internal):** Carried internally; unstreamlined internal-carriage version with parachute retarder; equipped with one 4-foot pilot parachute, one 16.5-foot ribbon extraction chute, one 64-foot solid chute and one 30-inch stabilization chute.

E. **B28RE (RE stands for Retarded External):** Carried externally; streamlined external-carriage version with parachute retarder; equipped with one 4-foot pilot chute and one 28-foot ribbon chute.

The **W28**—the warhead version was a lightweight, Class D thermonuclear warhead. It was used in the Hound Dog air-to-surface and Mace surface-to-surface missiles.

The Mark 28 bomb was involved in a couple of accidents. In 1966, a nuclear-armed *B-52* bomber and a KC-135 tanker aircraft collided during refueling over the village of Palomares in Spain. This collision caused the B-52 to break up and all the four bombs were ejected from the B-52. Of the four bombs, one landed intact, two of them detonated upon impact and the fourth bomb fell into the sea. In 1968, a B-52 carrying four B28 bombs was on a Cold War "Chrome Dome" alert mission over the Baffin Bay. It tried to make an emergency landing on the Thule Air Base after a fire broke out in the navigator's compartment, but instead it crashed onto the sea ice of North Star Bay, Greenland. This crash caused the nuclear payload to rupture and disperse, which resulted in radioactive contamination.

3.1.24 Mark 36—Heavy High-Yield, Two-Stage Bomb

The **Mark 36** bomb was a heavy high-yield, two-stage thermonuclear bomb. It was a more advanced version of the earlier Mark 21 nuclear bomb. It was developed in the 1950s during the same period as the Mark 15, Mark 17 and Mark 39. It was retired between 1961 and 1962. All Mark 21 bombs were converted to Mark 36 Y1 Mod 1 bombs in 1957. This bomb was replaced by the higher yield B41 nuclear bomb.

It had an air or ground contact burst option. Regardless of the burst option selected, it was always dropped in the parachute-retarded way. Because of its megaton yield, it utilized two parachutes to slow its descent, which in turn, allowed the aircraft more time to escape. It had two versions: the Y1 "dirty" version, which used a Depleted Uranium or U-238 tamper-pusher, and the Y2 "clean" version, which used an inert fusion stage tamper-pusher assembly such as tungsten or lead. The dirty version was associated with the 9-10 megatons yield, while the clean version would have a yield roughly half of that.

3.1.25 B39—Improved Version of the Mk-15

The **B39**, also known as the Mark 39, was a thermonuclear bomb. It was an improved and lightweight version of the MK-15. This bomb had better safety measures. It used a gas-boosted primary to reduce weight.

B39 entered service in 1957 and the last one was removed from service in 1966. It had a yield of 3.8 megatons and weighed 10,000 pounds. It had two variants. It was equipped with four parachutes for retarded delivery. It was delivered by B-47 and B-52 bombers. It was also supplied to Great Britain under Project "E" for the Valiant bomber. It also had a missile warhead variant, the **W39** warhead. The W39 was used on the Redstone intermediate-range ballistic missile, the SM-62 Snark intercontinental cruise missile and the B-58 Hustler weapon pod. The various versions of this warhead were Redstone Mk-39Y1 Mod 1 and Mk-39Y2 Mod 1, and Snark Mk-39Y1 Mod 1.

Both the B39 bomb and the W39 warhead used a common nuclear physics package which was derived from the Mark 15.

A B39 casing is on display in the Cold War Gallery of the National Museum of the United States Air Force in Dayton, Ohio.

In 1961, while on a 24-hour airborne alert mission, a B-52 Stratofortress broke up in mid-air and crashed near Goldsboro, North Carolina. During this accident, the plane dropped both the Mark 39 bombs it was carrying. One of the bombs was prevented from detonation and had a soft landing due to the proper functioning of its parachute. The other bomb's parachute experienced malfunction that caused it to fall free and break apart upon impact thereby creating a deep crater.

3.1.26 B41—Highest-Yield Weapon Ever Deployed, the Only Three-Stage Thermonuclear Device

The **B41** (designated as Mk-41 till 1968, from there on it was redesignated as B41) was the highest-yield nuclear weapon ever deployed by the Unites States Air Command in the 1960s. It was the only three-stage (fission-fusion-fission) thermonuclear weapon ever developed by the United States; often termed as the "thermonuclear triple threat." It was carried by the B-52 Stratofortress and B-47 Stratojet. It remains the highest yield-to-weight ratio of any weapon created.

Its development began in 1955 when the United States Air Force issued a requirement and a feasibility study for a Class "B" weapon (high-yield, over 10,000 pounds/4.545 kg). It was based on the Bassoon test device fired in the Redwing Zuni test in 1956; test firing of UCRL "clean" 3-stage concept for class "B" requirement. Then the Bassoon Prime device was fired in Redwing Tewa in 1956; test firing of UCRL "dirty" 3-stage concept for class "B" requirement. These tests were not the real test version of actual designs intended for deployment; they were conducted as a proof of concept. A lot of enhancements, redesign and testing were needed to meet the requirements and this is when Operation Hardtack Phase I came into picture. And then the feasibility study was completed and designation TX/XW-41 was assigned in 1956. Right through the period of 1958-1960, prototype tests were fired. The testing of clean variants of the TX-41 bomb, drop testing of the TX-41 ballistic shape, etc., was conducted. Thereafter, ordnance characteristics of the TX-41 were revised and accepted by the *Special Weapons Development Board (SWDB)* in 1958, and production engineering of the same followed soon. After this long process, the early production of the B41 bomb began in 1960.

The B41 had five fuzing options: it could be set for free fall air burst, free fall surface burst, parachute retarded air burst, parachute retarded surface burst, or laydown delivery.

It had two yields: less than 10 megatons and 25 megatons. It had "dirty" and "clean" configurations. Y1—a dirty version with a tertiary stage encased with U-238 and a 25 megatons yield, and Y2—a clean version with a lead-encased tertiary and a less than 10 megatons yield.

It was progressively replaced by the more versatile BK-53/MK-53. The first B41s were retired in 1963 and the last in 1976.

The **W41** was derived from the B41 and it was planned for use in the SM-64 Navajo cruise missile. The W41 program began in 1956, but was canceled in 1957.

3.1.27 B43—An Air-Dropped Nuclear Weapon

The **B43** was an air-dropped nuclear weapon. It was developed in 1956 by the Los Alamos National Laboratory and entered into production in 1959. It entered service in 1961, and these weapons were progressively phased out in the 1980s, and the last B43 weapons were retired in 1991 in favor of newer weapons. B43 was replaced by the B61 and B83 weapons.

It was a variable yield, low-drag, parachute retarded, shock-resistant nuclear bomb meant for high-speed, low-altitude laydown delivery. It was equipped with fuzing options for airburst, free fall, contact, ground burst or laydown delivery.

B43 was a Tsetse primary based design nuclear weapon (Tsetse primary was the basis of the core design of other weapons like the W44, W50 and W59 warheads, and the B57 bomb). It was built in two variants, Mod1 and

Mod2, each with five yield options. The dimensions and weight varied from version to version. The explosive yield varied from 70 kilotons to 1 megaton.

Several updated versions of the B43, Mods 2, 4 and 5, were introduced with new safety and delivery features. For training and practice purposes, several different types of practice bombs were produced, for example, BDU-6, BDU-8, BDU-18 and BDU-24.

B43 could be used by a wide variety of aircrafts: A1, A-3, A-4, A-5, A-6 and A-7. It was also supplied to Great Britain under Project "E" for the Canberra and Valiant aircraft.

This bomb is associated with the Broken Arrow nuclear accident that went unacknowledged by the Pentagon until 1981—*USS Ticonderoga* lost one of her A-4 Skyhawks off the coast of Japan that was carrying a B43 nuclear bomb on 5th December 1965 during a training exercise. No evidence of the aircraft, pilot or payload was ever found.

3.1.28 B46—Experimental Design which Evolved into the B53

The **B46** nuclear weapon (also called the Mark 46) was an experimental design, which later evolved into a new weapon design in 1959, the TX-53, which was redesignated the B53 nuclear bomb and W53 warhead. The B46 was intended to have a 9 megaton yield. It was tested but never deployed. Its deployment status was canceled in 1958.

B46 was designed and tested several times in the late 1950s. Originally intended to be a production design, the B46 ended up being only an intermediate prototype which was test fired several times. These experimental/test prototypes were known as TX-46 units. "Clean" and "dirty" versions were tested during Operation Hardtack I.

The **W46** warhead was derived from the B46 and it was planned for use in the SM-64 Navajo cruise missile. The W46 program was canceled in 1958 in favor of the Titan II/W53 combination.

3.1.29 B53—One of the Longest-Lived and Highest-Yield Weapon

The **B53** bomb has been one of the longest-lived and highest-yield nuclear weapons ever in the United States arsenal. Initially, it was designated as MK-53 and from 1968 onwards it was redesignated as B53. It was a high-yield bunker buster thermonuclear weapon.

National Nuclear Security Administration's (NNSA) Los Alamos National Laboratory and Sandia National Laboratories designed the B53 bomb. Development began in 1955. The B53 was based on the earlier Mk-21 and Mk-46 weapons. Two variants were built: the B53-Y1, a "dirty" weapon using a U-238-encased secondary, and the B53-Y2, a "clean" version with a non-fissile secondary casing.

The B53 was designed to hit targets deep underground. It is said to be 600 times more destructive than the atomic bomb dropped on Hiroshima in 1945. It weighed 10,000 pounds and was the size of a minivan.

The B53 was the basis of the **W53** warhead carried by the Mark 6 re-entry vehicle installed on the Titan II Missile. It was the highest yield (9 megatons) missile warhead ever deployed by the United States. The warhead used the same physics package as the B53. The warhead entered stock in 1962 and retired in 1988.

The B53 bomb was built during the Cuban missile crisis in 1962. It entered service in 1962 and retired in 1997. It played a key role in the United States' nuclear deterrent until its retirement in 1997. After nearly 50 years of service, the last remaining *B53* was dismantled at the Pantex plant in Texas in October 2011. In fact, the B53 dismantlement program, right from returning and staging it at NNSA's Pantex Plant to ultimate disposal, was safely and successfully completed a year ahead of schedule. This program is considered to be a major milestone towards achieving the vision of a world without nuclear weapons.

3.1.30 B57—Tactical Multipurpose Weapon

The **B57** nuclear bomb was a tactical multipurpose nuclear weapon developed by the United States during the Cold War to deal with the threat of Soviet submarines. It was a relatively small weapon in the inventory, and due its lightweight attribute, it could be carried by just about anything in the inventory. It was designed to be dropped from high-speed tactical aircraft. It was based on the Tsetse primary design. It entered service in 1963 as MK-57. After 1968, it was re-designated as B57. It retired from service in 1993. The depth bomb version of B57 replaced the United States Navy air-dropped nuclear depth bomb, the Mk101 Lulu.

B57 was equipped with a streamlined casing to withstand supersonic flight. It was intended for multipurpose usage—as a nuclear depth charge and a nuclear bomb to be used for land attack missions and anti-submarine warfare (ASW). It could be carried by a variety of carrier-based and land-based ASW aircraft like P-3 Orion, Nimrod MR2, SH-3 Sea King, A-6 Intrude, F-16 Fighting Falcon, and many maritime patrol aircraft. Under "dual key" arrangements, it could also be delivered by NATO allies.

It had six versions with yields ranging from 5 to 20 kilotons. The versions were: Mod 0, Mod 1, Mod 2, Mod 3, Mod 4 and Mod 5. Some versions had a parachute retarder to enable them to be dropped at low altitudes and give the carrier enough time to escape from the blast zone. It had various fuzing options and modes: free fall or parachute retarded, a hydrostatic fuze for use as a depth charge, and a radar *fuze* for airburst or ground burst. The fuzing options were selected on the ground prior to flight for ground or air burst. This device could be carried internally or externally.

3.1.31 B61— Lightweight, Flexible and Multipurpose

The **B61** is the primary thermonuclear weapon in the United States nuclear stockpile following the end of the Cold War. It is a family of weapons based on a single basic weapon and physics package design. It is extremely flexible, modular and multipurpose in nature. It is an intermediate yield strategic and tactical weapon. It has been deployed on a wide variety of tactical (F-100, F-104, F-4, F-105, F-15E, F-16, F-111, F-117 and Tornado) and strategic (B-52, FB-111, B-1 and B-2) aircrafts. It is perhaps the most versatile nuclear weapon in the United States stockpile. Approximately 400 bombs are scheduled to be used in quality and reliability testing through 2025.

The original B61 bomb has undergone many modifications, right from the initial development to the earth-penetrator version. It was developed by the Los Alamos National Laboratory starting in 1960. Initial manufacturing began in 1966 and quantity production began in 1967. The original models of B61 used PBX 9404 HMX based plastic bonded explosive (also known as PBX or polymer-bonded explosive), whereas newer models have triaminotrinitrobenzene based PBX, which is an insensitive high explosive that improves the munitions survivability and safety by preventing it from detonating due to shock, fire and impact.

B61 is designed as a lightweight weapon of intermediate yield that could be carried by modern high-performance combat aircraft. It has six basic modifications: Mod 0, Mod 1, Mod 2, Mod 3, Mod 4 and Mod 5. Major modifications were made to the B61 in 1966, 1975, 1977, 1979 and 1991. Its most recent modification is the B61 Mod 11 (B61-11)—an earth-penetrating nuclear device and it was introduced in 1997. B61-11 was designed to destroy underground military facilities and targets such as ballistic missile silos and command bunkers, which otherwise are indestructible targets from the air. The basic B61 bomb weighs approximately 700 pounds and the earth-penetrating version, the B61-11, weighs an additional 450 pounds. B61-11 is equipped with a special ground impact time delay feature to allow it to penetrate into the earth before exploding.

B61 has various fuzing options and modes: surface burst, air burst and subsurface burst. All the latest bombs in the inventory have been equipped with sophisticated security devices and full fuzing options. Moreover, the yield and fuzing options can also be configured electronically by the carrier aircraft; hence this bomb is also known as the "dial-a-yield-bomb."

Currently the project "BIOS" (Bomb Impact Optimization System) to develop a new weapon using the B61 warhead package is in progress. It is meant to be a glide-bomb adaptation of the B61 for aircraft delivery against defended target complexes. It has proposed using GPS (*Global Positioning System*) for guidance in order to achieve better accuracy.

The W69 missile warhead for the AGM-69 short range attack missile, the W73 missile warhead for the AGM-53 Condor air-to-ground missile, the W80-0 sea-launched cruise missile warhead, the W80-1 warhead for the air-launched cruise missile and the advanced cruise missile, the W84 for the Ground Launched cruise missile, and the W85 warhead for the Pershing II missile—all derived their basic design from the physics package of the B61.

The **W61** warhead was derived from the B61-7 laydown bomb. The W61 program was meant to equip the small Midgetman intercontinental ballistic missile with a strategic earth-penetrating warhead, but this program was canceled due to the cancelation of the Midgetman program.

3.1.32 B77—High-Yield Bomb Designed for the B-1A Bomber

The **B77** was a high-yield thermonuclear bomb designed to match the delivery capabilities of the B-1A bomber, and was also intended to be capable of delivery by a variety of aircraft besides the B-1A. It was required to have the ability to be dropped from supersonic speeds at altitudes of 60,000 feet, or in a laydown delivery at high subsonic speeds at altitudes as low as 100 feet. It was equipped with a roll control system that handled the various delivery modes. This system evolved from a cold gas version to a warm gas driven assembly. This system was responsible for fuzing and controlling weapon orientation prior to parachute deployment in order to ensure a stable flight. In the low-altitude laydown delivery mode, the roll control system maintained the upright orientation of the bomb prior to the parachute deployment. In the high-altitude free-fall delivery mode, it provided the torque required to spin the bomb for achieving stability. All these functions of the system depended on the gas generator output in terms of a constant output torque, minimized time for roll maneuver, etc.

This bomb was meant to replace the Mark 28 and Mark 43 bombs, but the B77 project was canceled in 1977. It was canceled because of the complexity, cost and excessive weight of the bomb delivery system, and also due to the cancelation of the bomber it had been designed to serve. It was replaced by the B83—many components of the B77 were used in the B83.

3.1.33 B83—The Most Powerful and Modern Strategic Bomb

The **B83** is a US-built high-yield gravity bomb with variable yield options—with a maximum yield of 1.2 megatons. It weighs 2,400 pounds. It is currently the most powerful and modern weapon in the United States arsenal and it will be retained indefinitely as part of the active stockpile. It may see a follow-on system that will replace the B83, while the bomb itself may be modified or re-roled.

B83 was derived from the B77 program as an affordable alternative to the costly B77 program. Much of the work undertaken for the B77 was transferred to the new program. Many components of the B77 were incorporated in the B83. The only the difference between the two is in the bomb body and delivery system. B83 replaced several earlier weapons like Mk-28, Mk-43 and Mk-53.

The development of the B83 design was carried out at the Lawrence Livermore National Laboratory in the late 1970s. The production engineering began in September 1980 and the first production bombs were completed by June 1983. Quantity production began in 1983 and it entered service in the same year itself, and the first underground test detonation took place on 15th December 1984.

This bomb introduced an improved low-level high-speed delivery capability. It has flexible fuzing and delivery options. It is intended for high-speed carriage and delivery from either high or low altitude. It is equipped with a Kevlar parachute that allows low-altitude delivery from a supersonic aircraft. It has the most modern arming, safety and security features, including insensitive high explosive main charge (will not detonate due to shock, impact or fire), trajectory sensing signal generator, enhanced electronic isolation, permissive action link (Category D PAL) locks and a fire-resistant pit. It is carried aboard strategic B-52 and B-2 bombers, as well as by tactical aircrafts like the F-16 Fighting Falcon, F-18 Hornet and F-18 Super Hornet, and AV-8 Harrier. It is a full fuzing option (FUFO) weapon as it is equipped with capabilities and options for retarded air burst, retarded ground burst or delayed ground burst (or laydown), and free-fall air or ground burst. The B83-1 is the only variant of the B83. The variant is a high-yield, low-drag device with a low-level laydown capability. It is equipped with improved features like a series of ring-like exterior teeth, a shock-absorbing nose, etc. All these features help to prevent the sliding of the weapon, minimize displacement on impact, etc.

B83 was considered as a potential candidate for the "Nuclear Bunker Buster" project, earlier known as the Robust Nuclear Earth Penetrator, but was subsequently canceled by the Congress. It has also been considered for use against any Near Earth Asteroids on collision course with Earth. For example, for handling the Apophis threat (Asteroid Apophis is a possible Earth bound asteroid), some scientists suggested deflecting Apophis off its course by delivering a volley of up to six nuclear missiles packing 1.2 megaton B83 warheads.

3.1.34 B90—Depth/Strike Bomb

The **B90** Nuclear Depth/Strike bomb was an American thermonuclear bomb designed in the mid-to-late 1980s for antisubmarine and land-attack warfare missions, which was subsequently canceled. It was intended to replace B57 bomb. It was scheduled for deployment in 1993, but it never entered military service as it was canceled in 1991. It had a design yield of 200 kilotons and weighed 780 pounds. It was intended for use as a land attack strike bomb, a nuclear depth bomb and a naval aircraft *weapon*. It was designed to be carried by A-6 Intruder and F/A-18 Hornet strike aircraft, S-3 Viking, P3-Orion, and new antisubmarine warfare aircraft, and by Dutch NP-3, Italian Atlantique and British Nimrod antisubmarine planes.

3.1.35 Pictographic Recap of American Nuclear Weapons

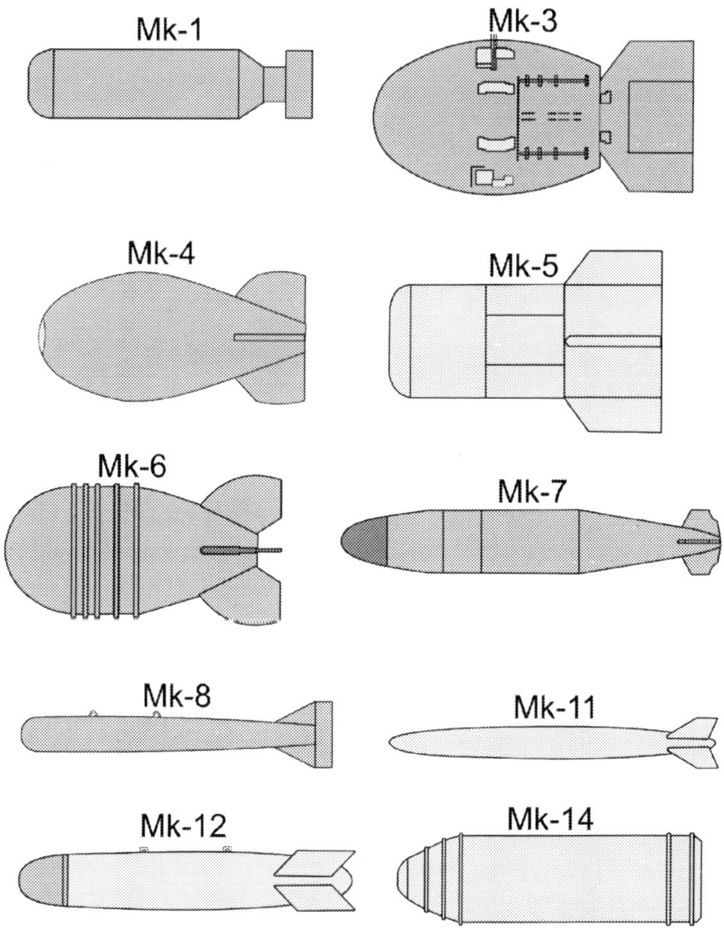

Figure 3.1: American Nuclear Weapons Part 1

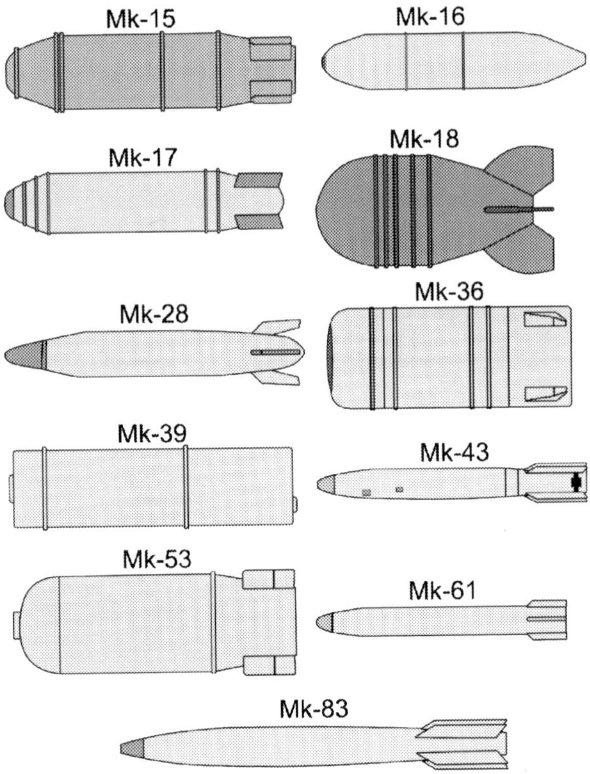

Figure 3.2: American Nuclear Weapons Part 2

3.2 Russia—The Second Country to Develop Nuclear Weapons

Russia was the second country to develop nuclear weapons. Four years after the United States dropped atomic bombs in 1945, the Soviet Union detonated its first nuclear device, RDS-1, also known as First Lightning, in the test codenamed "Joe-1," at the Semipalatinsk Test Site, Kazakhstan.

Russia is one of the five official nuclear weapons states under the Nuclear Nonproliferation Treaty. Today, it is believed to have the world's largest nuclear arsenal. Various estimates put Russia's stockpile, including reserves, at between 5,000 and 8,000 warheads. Russia's arsenal holds missiles launched from the sea or air, cruise missiles, anti-submarine weapons and torpedoes, and depth and gravity bombs.

After the dissolution of the Soviet Union in 1991, according to the terms of the Lisbon Protocol and Trilateral Agreement between Russia, Belarus and the United States, the nuclear weapons with Belarus, Kazakhstan and Ukraine were returned to Russia.

There have always been global concerns regarding the safety and security of nuclear weapons in storage facilities in Russia, and to add to this concern level is yet another factor—stolen or missing weapons. The United States is aiding Russia in enhancing the safety and security of nuclear facilities through initiatives like the Nunn-Lugar Cooperative Threat Reduction Program. International Science and Technology Center is another initiative through which countries like the Unites States and Japan are helping Russia with research and peaceful employment opportunities for nuclear scientists.

According to analysts, over the years, Russia's nuclear command and control system has become lax, but after the 2001 terrorist attacks, Russia increased security at nuclear weapons facilities and also invested in training, monitoring systems and maintenance programs. But there are still concerns regarding the same.

The latest development is that to put aside the tensions of recent years with the Unites States, Russia and the United States have signed the new START treaty with an aim to reduce nuclear arsenals on both sides by about one-third.

3.2.1 Tsar Bomba—Dubbed as the World's most Powerful Weapon

The **Tsar Bomba** (Emperor Bomb) is also referred to as the King of all Bombs and nicknamed 'Big Ivan' by its makers. It is dubbed as the world's most powerful weapon with an incredible yield of 50 megatons of TNT. It was developed in a short time span of only 14 weeks.

The bomb was originally designed to have a yield of about 100 megatons of TNT. Later, the bomb yield was reduced by half in order to limit the amount of nuclear fallout, and this was achieved by replacing a portion of the radioactive uranium with a lead tamper. It was officially designated RDS-220. It weighed over 27 tonnes. It derived the name Tsar Bomba from an analogy to two large objects—the largest bell in the world, the *Tsar Kolokol,* and the large, nearly 20-feet long cannon, the *Tsar Pushka.* The main aim of this weapon was to display USSR superiority in the cold war game and the arms race going on with the United States, and it did serve this purpose well by demonstrating USSR capability of producing extremely dangerous WMD.

It was a three-stage Teller–Ulam design hydrogen bomb, but the uranium fusion stage tamper of the tertiary (and possibly the secondary) stage(s) was replaced by one(s) made of lead.

On 1st September 1961, the testing moratorium that had been in place for almost three years was broken. The Soviet Premier, Nikita Khrushchev, had ordered the development of this weapon in July 1961. This project was initiated in July 1961. The bomb was tested over the Mityushikha Bay Nuclear Testing Range in the Novaya Zemlya archipelago in the Arctic Sea in October 1961, but it never entered service. The test created a nuclear explosion visible from over 610 miles away and released 3800 times more explosive energy than the Hiroshima bomb. This explosion was more powerful than all the explosives used by all countries during the World War II. The mushroom cloud had a diameter of about 40 kilometers and reached a height of 60 kilometers. About 80% of the fallout from the test was deposited as global fallout. Till date, this is the largest nuclear weapon detonated.

Though it was technically meant for aircraft-delivery, Tsar Bomba was too large to fit inside of the largest soviet bomber of that era, the Tu-95. Later on a Tu-95 was specially modified to carry this massive bomb to its test site. The bomb bay doors and fuselage fuel tanks were removed, the bomber was coated with a special white reflective paint to protect it from damage caused by the heat created from the detonation, and a fall-retardation parachute was attached to the bomb in order to slow the bomb's descent so that the aircraft could escape out of the blast zone. The immense yield and dimensions made the Tsar Bomba impractical for military use.

Other bombs like RDS-1, RDS-2, RDS-3, RDS-4, RDS-6, RDS-7, RDS-27, RDS-3I and RDS-37 had yields much lower in comparison to this extremely large bomb.

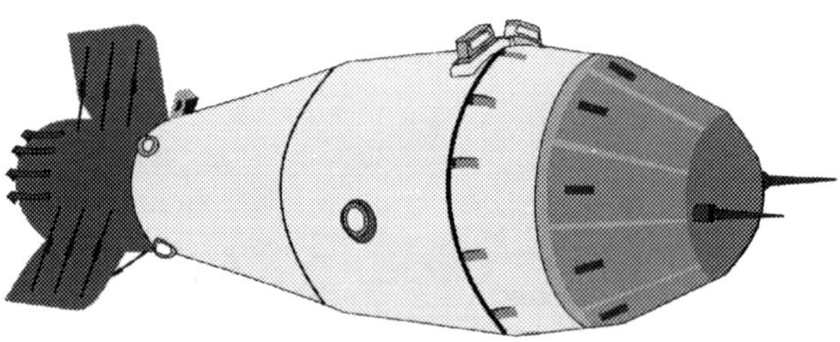

Figure 3.3: The Tsar Bomba

3.2.2 VA-111 Shkval—High-Speed Rocket-Propelled Torpedo

The **VA-111 Shkval** (squall)—a Russian-designed solid-rocket propelled underwater torpedo capable of a launch speed of 50 knots and a maximum speed of more than 200 knots underwater (this maximum speed is achieved when its liquid-fuel rocket ignites and propels it after the launch). This torpedo is four to five times faster than other conventional torpedoes. The extremely high speed is possible because of supercavitation. In supercavitation, what happens is while the weapon is on its trajectory under water, it is surrounded by a thin layer of gas that prevents water from coming in contact with its body surface, and which in turn significantly minimizes the drag factor and the metal-to-water friction. It is guided by an auto-pilot rather than by a homing head.

The NII-24 research institute was to develop a new weapon system capable of combating nuclear submarines. In the late 1960s, NII-24 merged with GSKB-47 to form the Research Institute of Applied Hydromechanics, and this merger later on produced this torpedo. It was designed as a rapid-reaction defense countermeasure against torpedoes launched by enemy submarines undetected by sonar and large ships such as aircraft carriers. The weapon can be fired either from a special container or a torpedo tube. This weapon was also known to be used as the PMK-2 deep sea mine. Its supersonic speed would give the target very little chance to carry out an evasion action. For this reason, it has been termed as a "revenge" weapon.

In 1995, it was publicly disclosed that Russia had developed an exceptionally high-speed unguided underwater missile way back in the 1960s. The first Shkvals entered service after a decade in development, back in 1977. It is believed to have entered service in the Delta I class.

VA-111 Shkval had a weight of 6,000 pounds, and had a length of 8.2 meters and a diameter of 533 millimeters. The older versions have a range of 2 kilometers whereas new versions have a range of 7 to 13 kilometers.

It has three variants: VA-111 Shkval is the original variant with an unguided system and a tactical nuclear warhead; Shkval 2 is the improved variant with a conventional warhead, a longer range and a guided targeting system; and Shkval-E is a shorter range and a less capable version marketed for export. The original design used a nuclear warhead, whereas all current designs are believed to use conventional explosives. Early versions had GOLIS autonomous inertial guidance system and upgraded models had terminal guidance system. In the inertial guidance system, once the torpedo is fired it moves in one straight line and that is why earlier designs had a shorter range and had to carry a small nuclear warhead. In order to add the dimensions of steering and maneuverability, a thrust vectoring system was introduced in the terminal guidance system to achieve accurate trajectory and to evade enemy surveillance.

Russia reportedly sold China 40 conventionally armed Shkval-Es in 1998, and this has definitely generated a wave of concern both in the Pentagon and in the world. Iran claims to have developed a new supercavitation torpedo, *Hoot*, which is apparently based on Russia's *VA-111 Shkval* torpedo. The arrest of a former U.S. Naval intelligence office, Edmond Pope, for espionage related to VA-111 Shkval design details, and the mysterious sinking of the famous Russian submarine, *Kursk*, are linked with this weapon.

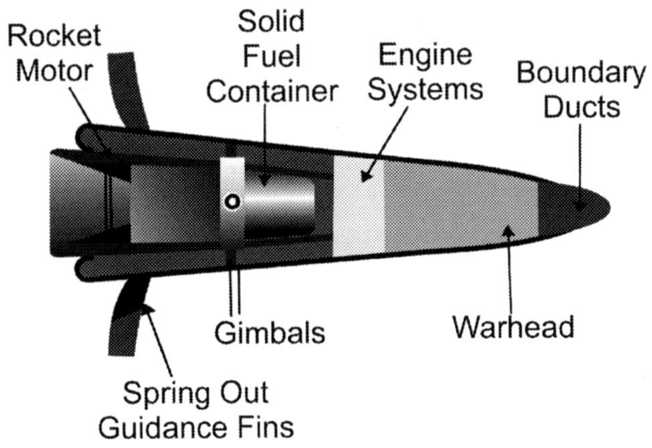

Figure 3.4: The VA-111 Shkval

3.3 United Kingdom—Has come a Long Way in Terms of Weapons, Research and Policies

The United Kingdom was the third country to test an independently developed nuclear weapon in October 1952. It is one of the five official nuclear weapons states under the Nuclear Nonproliferation Treaty. It has a strong track record in nuclear engineering and research. It was the first country to seriously study the feasibility of nuclear weapons. In recent times, the United Kingdom has also shared a different stand regarding tackling of new WMD scenarios like international terrorism. Several British experts believe that the major cause of such new concerns is overdependence on foreign fuels; they believe that by reducing dependence on foreign oil and gas, security can be increased considerably.

The United Kingdom had dismantled all its air-delivered nuclear weapons by 1998. Its nuclear forces are deployed on missiles carried on the Trident submarine fleet. The Trident program is the sea-based deterrent. Since the 1998 Strategic Defense Review, its nuclear deterrent is based entirely on submarine-launched missiles.

The United Kingdom has a cooperative attitude and approach to international security. It plays an important role through the 1958 USA-UK Mutual Defense Agreement. These countries have cooperated extensively on technological and operational matters, military cooperation and nuclear security.

Right from Blue Danube, the first operational nuclear weapon stockpiled to cold war weapons like Violet Club, and to modern-day weapons like Trident nuclear submarines, this country has come a long way, both in terms of its weapons and policies towards using them.

3.3.1 *Blue Peacock—A British Tactical Project*

The **Blue Peacock**, renamed from Blue Bunny and originally dubbed Brown Bunny, was the codename of a British tactical nuclear weapon project in the 1950s. It was the codename of a British 1950s Cold War defense project that aimed to place several 10-kiloton nuclear mines in parts of Germany in case of war, to thwart a Soviet attack by causing mass destruction and contamination over several kilometers to prevent subsequent enemy occupation. It was essentially a nuclear landmine which was conceived in the early days of nuclear deterrence.

Blue Peacock used the same physics package as that of the Blue Danube (Blue Danube was the first operational British nuclear weapon. It was based on the Hurricane device). The weapons were prototyped, developed and tested at the Armament Research and Development Establishment as part of a secret army "atomic demolition munitions" program. As the size of this device was large (more than 7 tons) and it had a bulky warhead, testing had to be done outdoors.

The 10 proposed devices were to have a plutonium core surrounded by a sphere of high explosives. They were to be detonated by a remote control or timer. Detonation could occur via a wire from an observation post up to three miles away, by an 8-day timer, or within 10 seconds if physically disturbed or damaged.

Technical challenges: scientists working on the project realized that the mines might not work after some days underground due to the electronics being too cold to operate properly. Various ideas to overcome this loophole were proposed and studied, such as wrapping the devices in insulating blankets or fiberglass pillows. One particularly remarkable (bizarre) proposal suggested that live chickens could be used to fill the casing of the nukes, and these chickens would then give off sufficient body heat to keep the delicate explosive mechanism from freezing, to keep all the relevant components at a working temperature—and for this reason the project was dubbed the chicken-powered nuclear bomb by the press.

In July 1957, the British Army decided to order 10 Blue Peacock mines to station them with the British Army in Germany, under the false pretext that they were atomic power units for troops in the field. Eventually, the project was canceled in 1958 over concerns and reservations about radioactive contamination and fallout hazard, and hiding nuclear weapons in an allied country was deemed as politically flawed. A lighter-weight ADM project called Violet Mist, based on the Red Beard tactical bomb continued in the place of Blue Peacock.

The only known prototype is now a part of the historical collection at the Atomic Weapons Establishment in the United Kingdom.

3.3.2 *Red Snow—A Result of US–UK Mutual Defense Agreement*

The **Red Snow** was a British thermonuclear weapon with a yield of 1.1 megaton. It derived its design based on the United States W28 nuclear warhead used in the B28 nuclear bomb and AGM-28 Hound Dog missile, differing in particular in the use of British high explosive around the primary. The Red Snow warhead had ballast added to maintain the overall weight, and ballistic and aerodynamic properties. It was intended to be carried in the Blue Streak (medium range ballistic missile), Blue Steel (an air-launched, rocket-propelled stand-off missile) and Yellow Sun Mark II (a free-fall bomb).

In 1958, the US–UK Mutual Defense Agreement was signed between the United States and United Kingdom on sharing information 'for mutual defense purposes.' The result was that the United Kingdom had access to American thermonuclear weapon design; one of the benefits of this agreement was the development of an anglicized warhead, Red Snow. It entered service in 1961 and was retired in 1972 as it was replaced by the WE.177 bomb (the WE.177 was Britain's last air-launched nuclear weapon, and it had three versions: WE.177A, WE.177B and WE.177C).

For a while in 1959, Red Snow was also associated with the Skybolt warhead requirement and with various schemes for tactical warheads with yields below 1 megaton.

3.4 France—Slow and Steady Yet Significant

French nuclear research began well before the World War II. It is one of the five official nuclear weapons states under the Nuclear Nonproliferation Treaty. Most of its weapons are believed to be deployed on submarines or on air-launched missiles. The relentless efforts and research pursuits of eminent physicists and scientists like Pierre Curie, Marie Curie, Frederic Joliot-Curie and Irene Joliot-Curie made a big difference in French nuclear research and development. But immediately after the war period, France lagged behind and was cut off from the rapid advances in this filed.

In order to speed up the research and development process, the *Atomic Energy and Alternative Energies Commission*, now *Commission for Atomic Energy and Alternative Energies* was set up in 1945. Over the years, it has conducted fundamental and applied research in various areas—nuclear research, technological research, extraction and production facilities, research centers for military applications, civilian research centers and several other facilities. Official approval for developing nuclear weapons was not authorized until late 1954. In 1954, the program "Bureau of General Studies" was created to develop an atomic bomb. The blow that France received in the Suez Crisis made it more determined to develop its own independent nuclear deterrent, and not depend on allies for support. Thereafter, in 1958, France's independent *Force de Frappe* (strike force) came into being. Through the early sixties, France concentrated on high-yield pure fission designs (AN-11, AN-22, MR-31) intended as strategic weapons. France began a program to develop ballistic missiles in September 1959. In 1965, a large plant at Pierrelatte started operations, initially producing only low enriched uranium, but in another two years time, by 1967 it made highly enriched uranium available for developing weapons.

In 1965, France began to shift its focus towards tactical weapons. During the 1960s, efforts were made towards the development of thermonuclear weapons. Thereafter, the 1970s was a period of several modernization programs and second-generation weapons, for example, extending the ability of aircrafts to deliver weapons by equipping them with a nuclear armed missile, EMP (EMP stands for electromagnetic pulse) hardening program, submarine fleet updating program, etc. Since 1992, France no longer produces weapons-grade plutonium. Since mid-1996, France has halted all production of fissile material for weapons.

3.4.1 *AN-11—First Nuclear Weapon Stockpiled by France*

The **AN-11** bomb was the first nuclear weapon stockpiled by France, entering service in 1964, to arm the Force de frappe (military strike force, especially the independent nuclear strike force of France). The bomb weighed 1500 kg and had an explosive yield of 60 kilotons. It was a pure fission weapon; it used plutonium as the only fissile material.

It was a free-fall bomb intended for high altitude delivery against strategic targets by France's first strategic nuclear bomber aircraft, the Mirage IVA.

Its development began in the late 1950s. It was stockpiled between 1963 and 1968. A development version of the weapon was fired in France's first nuclear test on 13th February 1960. The first AN-11 prototype was tested on 1st May 1962. A live drop from a Mirage IVA was conducted on 19th July 1966.

3.4.2 *AN-22—France's Second Air-Dropped Nuclear Weapon*

The **AN-22** bomb was France's second air-dropped nuclear weapon. It replaced the earlier AN-11. It was based on the A-11, but also had enhanced safety features and was equipped with a retarding parachute mechanism for low-level delivery. It was a pure fission plutonium bomb intended for free-fall delivery from Mirage IVA bombers.

Initially, the AN-22 weighed 1400-1500 kg with a yield of 60-70 kilotons. Over time, modifications like redesign of the casing were made whereby almost 50 percent reduction in weight was achieved while maintaining the yield.

It entered into service in 1967 and retired in 1988. As the Mirage IVAs were retired in the late eighties so were their bombs. The last was retired in 1988 in favor of the Air-Sol Moyenne Portée (ASMP is a French air-launched nuclear missile) stand-off weapon.

3.4.3 AN-51—French Tactical Warhead for the Pluton Missile

The **AN-51** was a French tactical nuclear warhead used on the Pluton short-range missile (Pluton missile was used as a French nuclear-armed short-range ballistic missile (SRBM) system during the Cold War). This weapon was relatively lightweight; it weighed about 500 kg.

The AN-51 Pluton warhead was stockpiled from 1973 to 1993. It was deployed with the Pluton missile in 1974.

The AN-51 was based on a pure plutonium fission warhead design called the MR-50 *common tactical charge* (*CTC*, charge tactique commune). It had two yield variants—one with a low yield of 10 kilotons and a high yield version of 25 kilotons.

3.4.4 AN-52—Yet Another French Tactical Warhead

The **AN-52** was a French tactical nuclear weapon. In contrast to the earlier AN-11 and AN-22 strategic bombs, the AN-52 was intended for tactical purposes only. It was based on the warhead design–the MR-50 *common tactical charge* (*CTC*, charge tactique commune); same as that of AN-51. There were two yield variants–one with a low yield of 6-8 kilotons and a high yield version of 25 kilotons.

It entered service in 1972 and was retired in 1992 in favor of the ASMP missile. A total of 80-100 were built between 1972 and 1982. The French Air Force and Navy used the AN-52 in the 1970s and 1980s. It was deployed on the Dassault Mirage IIIE and SEPECAT Jaguar A aircraft of the Air Force in 1973 and 1974 respectively, and on the Dassault Super Etendard for naval aviation in 1981.

3.4.5 MR-31—Pure Plutonium Device

The **MR-31** was a French-built missile warhead, which entered service in 1970 and was withdrawn from service in 1980. It was probably the highest yield plutonium fission device ever developed, and for this reason it was difficult to harden it against pre-detonation effects.

It was meant to arm the SSBS S2 (*Sol-Sol Balistique Strategique*) intermediate-range ballistic missile. It remained in service until the last SSBS S2 was retired; the S2/MR-31 combination was replaced by the SSBS S3/TN-61.

It was a pure plutonium device that weighed 700 kg and had a yield of 120 kilotons.

3.4.6 *MR-41—France's First Boosted Fission Warhead Based on HEU*

The **MR-41** was France's first boosted fission warhead based on highly enriched uranium combined with deuterium and tritium, with a yield of approximately 500 kilotons and a weight of 700 kg. Encelade was a proof test of the MR-41 boosted fission warhead conducted in 1971. It was stockpiled from 1971 to 1979.

It was meant to arm the M1 and M2 missiles in *Redoutable* class ballistic missile submarines. The initial development of the warhead began in 1963 and a second phase ran from 1966 to 1971. It went into operational service with the first patrol of Le Redoubtable in January 1972. It was in service from 1972 to 1979, and was replaced by the TN-60.

3.4.7 *TN-60 and TN-61—A Family of Thermonuclear Warheads*

The **TN-60/61** was a family of thermonuclear warheads. The development process was a lengthy one spread over 8 years and 21 tests. The development began in 1968. Both the warheads were used on the MSBS M20 and MSBS M4 (submarine-launched missiles), and the SSBS S3 (land-based missiles).

The **TN-60** was France's first thermonuclear weapon. It was a one megaton missile warhead. It entered service in early 1977 and was replaced with the TN-61 starting in late 1977.

The **TN-61** was a lighter weight version of the TN-60. It was hardened against nuclear weapon effects like electromagnetic pulse (EMP). It was in service between 1977 and 1991.

3.4.8 *TN-70 and TN-71—Lower Weight, Lower Yield and Higher Survivability*

The **TN-70/71** family of thermonuclear warheads had a lower weight, lower yield, and higher survivability in comparison to its predecessors, TN-60 and TN-61. These warheads were miniaturized and extremely hardened that made them less vulnerable to EMP effects. The smaller warhead size made it possible to use TN-70/71 for arming missiles with multiple warheads.

The **TN-70**, a French-built thermonuclear warhead, which was used on submarine-launched ballistic missiles in Redoutable class ballistic missile submarines. It entered service in 1985, and was then replaced by a better version TN-71 in 1987, and finally withdrawn from service in 1997.

The **TN-71** had a lower radar cross section and was more survivable against ABM defenses. It was much lighter in weight; a reduction in weight of 25 kg. It entered service in 1985, was then replaced by its miniaturized, hardened and stealthy successor, TN-75. TN-71 was withdrawn from service in 2004.

3.4.9 TN-75—Strategic Warhead at par with the United States

The program for the **TN-75** warhead began in 1987 at the *Centre d'Etudes de Valdue* and it was tested numerous times before 1991, when the development testing ended. This warhead brings French strategic warhead technology up to par with the United States; experts termed it to be a major technological leap. It is a France-built stealth, miniaturized, safety-enhanced, hardened MIRV thermonuclear warhead with a yield of 100 kilotons. It is the successor of TN-71.

It was used on France's M45 submarine-launched ballistic missiles, carried by the last of the *Redoutable* class submarines, S616 Inflexible, and by the *Triomphant* class submarines.

It had a low radar cross section to evade detection and interception. Its enhanced, precisely computed shape and new coating material gave it better penetration during reentry. It was also equipped with more decoys to divert anti-ballistic missile defenses.

The former French President, Jacques Chirac, issued a statement in 1995, which said, that a final certification test of TN-75 would be needed before its deployment and series production. It is likely that its first full-scale test with a reported yield of 110 kilotons was conducted on 1st October 1995 at Fangataufa. Thereafter, series production of this warhead began at the Centre d'Etudes de Valduc.

3.4.10 TN-80 and TN-81—Warheads for the ASMP Platform

The **TN-80** was a miniaturized, hardened nuclear warhead for the Air-Sol Moyenne Portée (ASMP) medium-range air-to-surface missile carried by the Dassault Mirage IVP bomber (the *Dassault Mirage IV* was a French jet-propelled supersonic strategic *bomber* and deep-reconnaissance aircraft). TN-80 became operational in 1985; it was deployed between 1985 and 1991. It was subsequently replaced by an enhanced version, the TN-81.

The ***TN-81*** was a miniaturized and hardened nuclear warhead designed specifically for the ASMP platform. The development of the TN-81 commenced in 1974. The improvised version was first tested in 1984 and went into production in 1985. It entered service in 1988 on the Mirage 2000N platform of the French Air Force, was then deployed on the Super Etendard of Naval Aviation, and finally replaced the TN-80 on the Dassault Mirage IVP bomber in 1991.

The TN-80 and TN-81 warhead family is similar to the TN-70/71 in technical sophistication. The TN-80 and TN-81 had a yield of 300 kilotons and a weight of about 200 kg.

3.4.11 TN-90—French-Built Variable-Yield Warhead

The **TN-90** was a variable-yield thermonuclear warhead with a maximum yield of 80 kilotons. It was developed by France for the Hadès tactile missile (the Hadès system was a short-range ballistic pre-strategic nuclear weapon system designed by France). In the late 1980s and the early 1990s, tests for this warhead were conducted.

Its development began in 1983, series production began in 1990 and it entered service in 1992. It was never actually deployed to the field. With the retirement of all French land-based missiles in 1996, both the 30 Hadès missiles and their TN-90 warheads were transferred to storage at Valduc, and eventually retired.

3.5 Questions for Reflection

A. What penalties do you think countries should face if they attack others using nuclear weapons?
B. How do you think your country would respond to a ban on the production of new nuclear weapons?
C. What additional provisions or treaties need to be implemented to deal with the spread of nuclear weapons?
D. How did nuclear weapons impact the Cold War?
E. What are the unique challenges of today's nuclear age?
F. What threat does the world face from countries with undeclared or unconfirmed nuclear weapons?

PART III

Radiological Weapons—The Global Threat Revisited

4

Radiological Weapons—Weapons of Mass Disruption

The 'R' in the acronym CBRN stands for radiological weapons. A radiological weapon (or radiological dispersion device, RDD) is a weapon that uses conventional explosives to disperse radioactive materials such as Cobalt-60 (Co-60), Cesium-137 (Cs-137) or Iridium-192 (Ir-192). These weapons can also cause damage and casualty by fire, or by spreading radioactive contamination through the water resources (dilution through water contamination, for example, contaminated wells or reservoirs) and food chain (food poisoning), or by air (for example, inhalation of dust or smoke). Radiological weapons can be used to contaminate critical areas and setups, attack troops, destroy livestock and food, make evacuated areas uninhabitable, and cause panic and causalities in densely populated civilian areas.

The radiological source, sophistication and size of the conventional bomb, availability, half-life, energy, quality and type of radiation, portability, longevity, size of the explosives, physical and chemical form of the radioactive material, and the environment and weather conditions (wind and precipitation) greatly impact the effectiveness of radiological weapons in terms of the size of the affected area and the level of destruction caused.

SPOTLIGHT

Though radiological weapons are not capable of causing a high number of casualties, still the economic instability and psychological impact caused by them could be severe and difficult to deal with.

Unlike nuclear weapons, radiological weapons are not considered as typical weapons for military purposes, but their 'silent' nature in terms of the fact that some radio-isotopes remain active for years or even decades after the contamination render the affected areas and facilities useless for a long time, disrupt life and paralyze economy—this is enough to create chaos, panic and fear. They could also be used to force mass evacuation. And this is why they are also called as "weapons of mass disruption." These are all offensive uses. Examples of defensive attack can include denial of a territory to an enemy by contaminating it thereby making it either impassable or uninhabitable.

Several attempted uses of these weapons and thefts of radioactive materials have occurred in the past, for example, a cache of radioactive cesium was discovered in Ismailovsky Park in Moscow (1995), unidentified thieves attempted to steal a container of radioactive materials from the Radon Special Combine chemical factory in Chechnya (1998), etc.

4.1 Dirty Bomb—Conventional Explosives Packaged with Radioactive Materials

A dirty bomb is a conventional explosive packaged with radioactive materials, which when exploded scatters radioactive dust, smoke or other material in order to cause radioactive contamination; hence the attribute "dirty." It is a combination of explosives such as dynamite and radioactive powder or pellets. A dirty bomb does not require a chain reaction like fission or fusion.

Designing this bomb is not much difficult; the hardest part is acquiring the radioactive material. Since the 9/11 tragedy, the fear of terrorist groups and rogue scientists using dirty bombs has increased significantly. The relative ease of assembling a dirty bomb makes it the cause of an ongoing debate among experts regarding its usage for inflicting damage by terrorist groups. Again the effective use of dirty bombs by terrorists depends on whether or not they have access to the sophisticated technologies needed to work with radioactive materials. Moreover, gamma emitting material is so radioactive that it cannot be deployed without covering it with a shielding material, which would make transporting it by plane or car impossible as it becomes too heavy to maneuver, and at same time, a lot less effective if detonated. Terrorist groups designing an effective device is also doubted due to the complex chemical design involved. Moreover, they should be able to lay their hands on sources that are sufficiently radioactive and dispersible to effectively contaminate and damage an area. All these factors make a dirty bomb a "hi-tech" weapon.

A WORD OF CAUTION

With a dirty bomb, terrorists can force people to abandon the places where they live and work. With a dirty bomb, they can force people to do it themselves on a much larger scale due to the psychological fear associated with this weapon.

A dirty bomb is considered as a weapon of mass disruption, not from the explosion perspective, but more from the fear and panic it creates in people for its radioactivity effects, evacuation measures and the after-effects like disruption of facilities and a paralyzed economy.

A dirty bomb has made a name for itself in popular culture, movies, documentaries, television series and books.

The immediate effects are limited to those resulting from the conventional explosive used. The radioactive material and radiation from the bomb will not be that effective in killing people, but the psychological impact and chaos of the effects caused by a dirty bomb have a much higher gravity as radiation exposure is considered to be an unknown source of danger because it is something that cannot be tasted, seen or felt. The 'real' danger becomes apparent much later after the explosion.

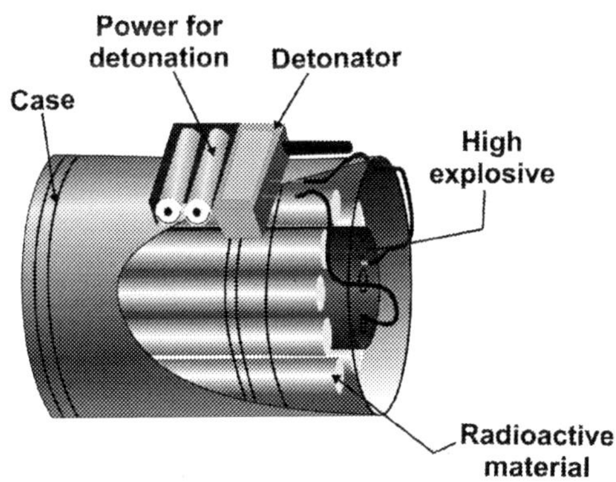

Figure 4.1: Generic Design of a Dirty Bomb

There are no incidents of a real dirty bomb ever detonating, however, there have been a few close calls and worrisome instances in the last couple of years involving black-marketed, orphan or loose sources, which indicate that a dirty bomb danger is real and is no longer a remote possibility. For example, according to intelligence reports, Iraq tested a crude radiological device in 1987. In another incident, in 1998, nineteen tubes of cesium went missing from the Moses Cone Memorial Hospital, Greensboro, North Carolina. In a recent incident in New Delhi, India, cobalt-60 was found in a scrap metal market, and the National Disaster Management Authority *(NDMA is the apex institution for disaster management in India)* identified six sources of cobalt-60 from this scrap market. Even unintended accidents indicate a dirty bomb scenario of a lethal combination of radiation damage, psychological impact and explosive destruction, for example, the explosion of the *Chernobyl Nuclear Power Plant* is widely considered to be the worst nuclear power plant accident in history. This plant exploded on 26th April 1986 during a routine maintenance and test schedule. During the shutdown and test routine, the reactor went out of control, which resulted in high radiation levels. This disaster not only had an impact on Russia, Belarus and Ukraine, but also created panic and fear within other countries, which lingers on even today.

4.2 Salted Bomb—The Doomsday Bomb

A salted bomb is also referred to as the "Doomsday" bomb or a "Weapon of Total Destruction." Surround a nuclear weapon with highly radioactive materials like gold or cobalt and you get a salted bomb. The yield from a salted weapon is usually lower than from a nuclear weapon of similar size. Salted versions of both fusion and fission weapons can be constructed.

The main purpose of a salted bomb is to not only produce a maximum physical effect but also to create a cloud of lethal, long-lived radiation over the targeted area.

A salted bomb is a variation of a nuclear weapon designed to produce enhanced quantities of radioactive fallout. It is simply an atomic or hydrogen bomb to which a salting or doping of a radioactive material is added. In other words, it is similar to the fission-fusion-fission weapons, but instead of a fissionable jacket around the secondary stage fusion fuel, a non-fissionable cover of a salting isotope (strontium-90, cobalt-60) is used, for example, in a cobalt bomb, cobalt-60 is used. This cover or blanket helps to capture the escaping fusion neutrons in order to maximize the fallout hazard. The weapon's tamper would be composed of ordinary cobalt, which the nuclear explosion would then convert to the radioactive isotope, cobalt-60. Instead of generating additional explosive force from the fission process, cobalt is transmuted into cobalt-60. *Leo Szilard,* a Hungarian American physicist and inventor, originated the idea of a cobalt bomb in the 1950s.

Variable fallout effects can be obtained by using different salting isotopes. The main proposed salting agents are gold, zinc, tantalum and cobalt. Some isotopes are more radioactive and some are insufficiently radioactive; the half-life factor determines this aspect. Isotopes with an ideal half-life like cobalt-60 are more stable as they will not cause decay before the fallout settles thereby making it useless to hide in shelters. Gold has been proposed for short-term fallout (days), zinc and tantalum for fallout of intermediate timeframe (months), and cobalt can be used for long-term duration (years). The effects depend on the half-life of the isotope used.

No salted bomb has ever been atmospherically tested. Nobody has ever built or admitted to building a salted bomb. The United Kingdom did test a bomb that incorporated cobalt as an experimental radiochemical tracer at its Tadje test site, Maralinga range, Australia, in 1957. This experimental explosion failed and was never repeated.

4.3 Delivery Systems—From Passive to Active Means

A "dirty bomb" is one possible way of dispersing radioactive materials. It is not a true nuclear weapon (no nuclear fission occurs) and does not yield the same destructive force as that of a nuclear weapon. Another radioactive source is RED—radiation emitting device. It is a silent weapon that emits radiation while concealed. It could prove to be a real danger as it could take a long time before a silent irradiator is discovered since there is no explosion or any other sign of damage. By the time it is discovered, several people could already have symptomatic radiation effects. Radiation poisoning by ingestion or inhalation is yet another source of danger; it is generally associated with very large doses of ionizing radiation, which can occur as a single large exposure (acute), or a series of small exposures spread over time (chronic).

Radiological weapons can be delivered through active means of (non-nuclear) explosives or through passive means (aerosol) of aerial dusting. They could be fabricated for delivery by cruise or ballistic missiles, or even as aircraft-delivered bombs. They can be in either solid or liquid form and delivered by aerosolized means. Radioactive materials can be delivered in various other ways too—destruction of a facility, plant or vessel containing radioactive material, through a human agent, by placing a container of radioactive material in a public place, or by contamination of water resources.

A WORD OF CAUTION

Terrorist groups could launch an attack on a nuclear power plant reactor so that it acts as a radiological weapon. Insecure nuclear facilities throughout the world compound this problem.

Radiological weapons may well be within terrorists' reach, which they can use to create panic in people and contaminate a large area. They could also render a great deal of property useless for an extended time period, unless costly remediation is undertaken to thoroughly clean the contaminated areas. Past incidents of smuggling and black marketing indicate that it is far easier to obtain some form of radioactive materials than fissile materials. Economic hubs like ports or densely populated areas are potential targets of a radiological attack by terrorists.

4.4 Materials—Industrial Sources, Medical Facilities

The materials/sources are readily available and do not require sophisticated or expensive delivery vehicles.

Any nuclear material like nuclear waste or medical isotopes is needed. Radioactive isotopes with extremely short half-lives are unsuitable for use in RDDs and REDs. The radiological weapons could be assembled by using powdered Polonium-210 or an emplaced radiological source such as Cs-137, Co-60, Americium-241, Californium-252, Iridium-192, Plutonium-238, Radium-226 and Strontium-90. Cs-137 is the likely weapon of choice as it is readily available at food processing plants, research applications and medical facilities. Radioactive materials are available from a wide range of relatively non-secure facilities. Some of the sources are nuclear power plant fuel rods, food and blood irradiators, medical devices, industrial plants, radiopharmaceuticals and research laboratories.

It is estimated that approximately one source is lost every day of the year due to which there are thousands of such "orphan" sources. 20 percent of these sources are classified as a potential high security concern when it comes to building radiological weapons. Russia is one such country that is believed to have a large number of such orphan sources. One of the most dangerous radiological incidents involving an orphan source was that of the Goiânia accident. In this accident, an abandoned radiotherapy source from an abandoned clinic was stolen and it resulted in external and internal contamination, and even death of many due to exposure.

4.5 Effects and Responses—From Panic to Peace

Trauma is the one of the main effects. Dealing with public fear towards radiological weapons may prove to be the greatest challenge due to the lack of public awareness regarding radiological weapons. And this public fear and lack of awareness affect the cleanup procedures in terms of time and money. Even after the cleanup procedures are completed and the radiation is reduced to an acceptable level, there might be public fear and unwillingness to return to the site. This public overreaction can also lead to more deaths caused by accidents and pandemonium during fleeing.

The effects are similar to the residual radiological effects of a nuclear attack. The effects of a radiological weapon will depend on whether the exposure is "chronic" (i.e., extended) or "acute" (i.e., brief). Factors like the type and amount of radiation absorbed, distance from the source of radiation and the duration of exposure determine the effects on people. The effects can include skin burns, nausea, loss of organ function, etc., and in longer exposure periods, serious diseases like cancer can occur.

Gamma-radiation emitters may cause whole-body irradiation. Three types of radiation-induced injuries can occur: external irradiation, incorporation of radioactive material into body cells, organs or tissues, and contamination with radioactive materials.

In the case of a concealed RED, people are not aware that they are being exposed to radiation until they begin to experience symptoms.

A dirty bomb would boost the radiation level above the normal level thereby increasing the risk of cancer and radiation sickness, which wouldn't cause instant death but would kill people slowly over the years—something like slow poisoning.

If radiological attacks were to occur, the speed of evacuation dictates the level of human exposure and response effectiveness in terms of radiation monitoring, contamination control, medical services and decontamination. The first response is that the affected people need to move away from the immediate vicinity of the blast as soon as possible. The local authorities need to immediately cordon off the affected areas.

SPOTLIGHT

A comprehensive plan for educating the public about radiation effects and radiological weapons is necessary to bring about a speedy and effective emergency response.

A large enough stockpile of drugs to treat victims of a radiological attack needs to be maintained. The emergency response plan for potential areas that can be targeted needs to be clear and comprehensive. It needs to include public education campaigns to understand the realities and myths of such attacks. Topics such as immediate and long-term health issues of exposure, shielding measures for protection, protective clothing and equipment, etc., need to be a part of the training sessions.

Communities of local professionals with the knowledge and practical training required to quickly address health hazards need to be in place. Authorities and counterterrorism units need to realistically appraise radiation risks before tackling such attacks. They also need to conduct education programs, training exercises and drills for people (civilians, media personnel, soldiers and medical practitioners) to prepare them with the necessary skills needed for dealing with radiological accidents and attacks.

4.6 Questions for Reflection

A. What is radiological terrorism?

B. Is there an international treaty that bans the manufacture of radiological weapons?

C. Is there an international body that regulates radioactive materials?

D. Why radiological weapons are a more likely candidate for terrorist attacks over nuclear, biological and chemical weapons?
E. How does radiation affect the public?
F. What happens with the radioactive waste from nuclear power plants?
G. Has a dirty bomb ever been detonated?
H. What are the terrible possibilities of a Doomsday bomb?

PART IV

Biological Weapons—The Malignant Microworld

5

Biological Weapons—Biowarfare, Bioterrorism and More

Several countries had biological weapons programs in the past, but many of them have renounced their biological warfare projects as they are now a member of the Biological and Toxin Weapons Convention. For example, the United States had a biological weapons program from 1943 until 1969, but has dismantled it since the country has become a member of the convention. Another example is that of Libya. Despite this convention, the development of biological warfare weapons has continued. According to controversial reports and evidence, more than a dozen countries are suspected to have continuing offensive biological warfare programs. Biological warfare has come a long way, right from the World War I to the 2001 anthrax mail attacks. All these instances indicate that the weaponization, research and use of biological agents will continue in a covert manner despite conventions and prohibitions.

A WORD OF CAUTION

Many constraints that previously limited terrorist use of biological agents are eroding. This alarming trend is directly proportional to the growing number of potential terrorist attacks; terrorists could develop new strains of biological agents that are resistant to treatment.

Several tests were conducted by the United States between 1964 and 1968. These tests showed that a single weapon is able to cover 2400 square kilometers with 30% casualties.

The rise of modern biotechnology and genetic engineering has now made it possible to produce classical biological agents much easier than it was 25-30 years ago. In modern times, biological warfare poses a great threat to our biodiversity—animals, food and water, humans, vegetation, etc. Since letters with weaponized anthrax spores were sent to the American Senate and media outlets, bioterrorism has gained attention not only in the United States but has become a worldwide concern too.

5.1 Delivery Systems—From Basic to Advanced

Biological weapons consist of four main components: biological agent, delivery system, container and a dispersal system. And the effectiveness of all these components depends on the type of agent and its stability, incubation period (the time elapsed between exposure until onset of signs and symptoms), virulence and lethality, physical properties, method of dissemination and weather conditions like temperature, humidity, wind speed and precipitation. Weather conditions play a deciding role in determining the effectiveness of bioweapons. For example, on a sunny day, the effectiveness will be minimized as there is a lot of ultra violet radiation that would deactivate pathogen agents.

Weaponized organisms can be used in their naturally occurring state or they can be modified to increase their stability and virulence.

Biological weapons can be used to contaminate livestock, water and food supplies, which in turn, when ingested can cause harm to humans.

They can be delivered in the aerosolized form; they can be dispersed in the air by using dry or wet aerosol sprayers. In the dry form, the agents are more stable, while in the wet form, they are less stable. Once inhaled, they can enter the bloodstream easily because of their small size and they also replicate themselves once disseminated.

They can be dispersed through mechanical devices like crop dusting aircraft or through explosive devices like bio-bombs. The delivery methods can range right from a sprayer mounted on a truck and a spray generator under the wing of a jet to including the agents in objects like envelopes, and from spray tanks and bomblets to a specialized cluster warhead carried on a missile.

Modern-day delivery methods—Remote control devices are a potential means of delivery, for example, agents could be placed in a container attached to an aerosol-producing pump that could be activated by a timer or a remote control. A cruise missile or an unmanned aerial vehicle could also be modified to spray its tanks with a bioweapon. Bioweapons could also be delivered by a human, for example, a suicide bomber could infect himself and then spread this infection by detonating himself in a crowded place like planes, malls, subways, etc.

5.2 Biological Agents—Potential Warfare Weapons

A biological agent is also called a bio-agent or biological threat agent. It can be classified into living organisms, pathogens and toxins. Pathogens include bacteria, rickettsiae, viruses and fungi. Toxins are chemical by-products of biological organisms. Only a small number of biological agents have the necessary characteristics to make them potential weapons (as the rest are either rare or cannot survive outside narrow temperature ranges).

The common modes of exposure or contracting biological diseases/ infections are inhalation, cutaneous contact (wounds or cuts), respiratory droplets and ingestion. We will discuss these modes in the subsequent subsections.

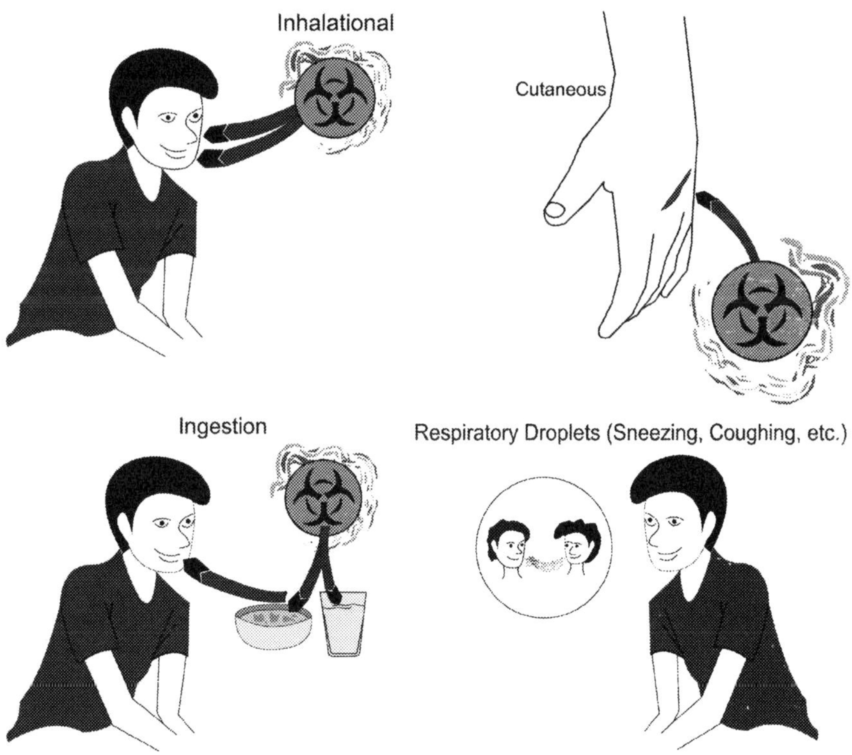

Figure 5.1: Common Modes of Exposure

Diseases caused by biological agents, both in their natural occurring state or in weaponized form, are either infectious or contagious. Not all infectious diseases are contagious, whereas some infectious diseases spread from one person to another, such infectious diseases are said to be contagious too.

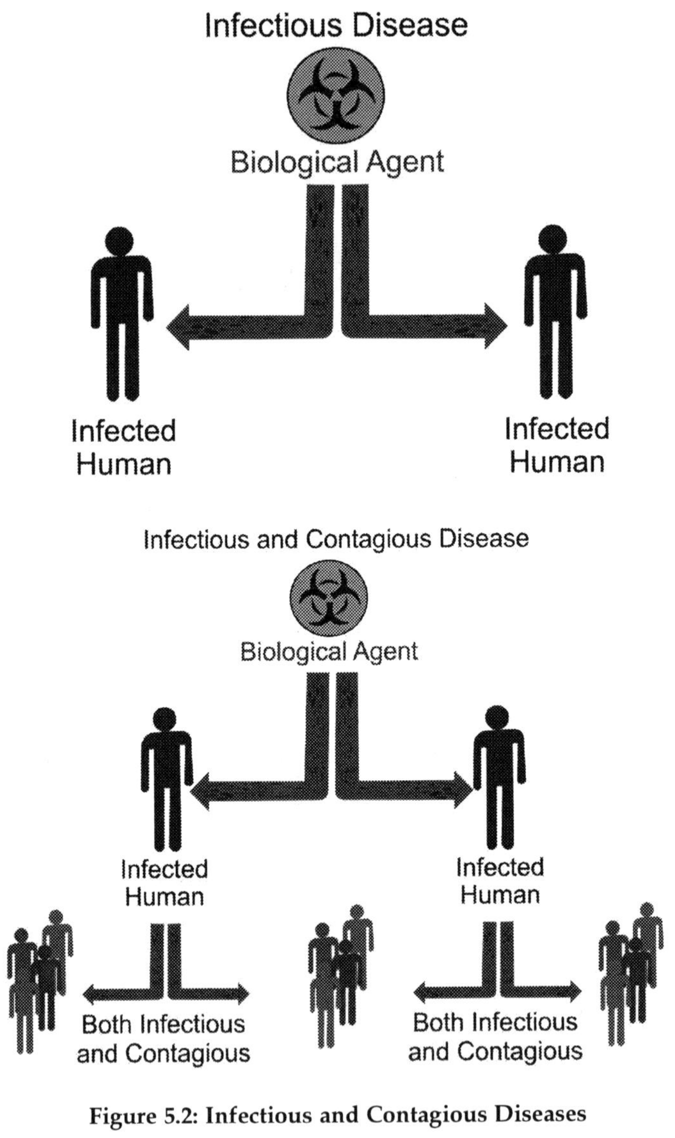

Figure 5.2: Infectious and Contagious Diseases

The Centers for Disease Control and Prevention (CDC) has categorized biological agents that can be weaponized by their relative level of risk. There are three categories of biological weapons. Anthrax, botulinum toxin, ricin, plague, smallpox and tularemia are on the top of the CDC's list of biological weapons. They are considered as "Category A" weapons as they carry the highest priority and risk level; they are most likely to be used in an attack, are highly lethal, cause public panic, strain the public health care system, and also much of the biodefense research effort is directed at these *agents*. "Category B" agents have the second highest priority, which cause low death rates and are considered less lethal. "Category C" agents carry the third highest priority and include emerging pathogens that could be engineered for mass dissemination and could have the potential for bioterrorism acts in the future.

Biological agents when used for biological warfare can be difficult to detect when released and the recognition of agent may take weeks. As they are easy to hide and difficult to detect, biological weapons pose a significant threat as WMD. As the field of biotechnology advances and with the increasing availability of dual-use know-how, the means to make agents more lethal and sophisticated will also progress.

Biological and chemical weapons differ in terms of their dissemination, makeup, scientific properties, origin and effects. The biggest difference between biological weapons and chemical weapons is that biological weapons are living and naturally created diseases and chemical weapons are artificially made poisons and toxins. Apart from this difference, biological weapons have the potential to spread over a large area and they can be contagious in nature, whereas chemical weapons only affect people in the affected area. Biological weapons have a long incubation period in comparison to chemical weapons. The biggest similarity between the two is that they both are weapons of mass destruction.

5.3 Tabular Recap of Bioweapons

Here is a quick tabular recap of the various biowarfare and bioterrorism agents. In the following subsections, we will discuss in details the potential weapons of bioterrorism and biological warfare. (Table 5.1)

5.3.1 1918 Influenza A (H1N1)—The Mother of all Pandemics

A. Introduction

The ***1918 Influenza*** pandemic (a pandemic is a widespread epidemic that affects countries or the entire world), also called the *La Grippe* or *Spanish Flu*, was a severe viral flu that infected one-third of the world's population, resulting in the deaths of 10% to 20% of those who were infected (3% to 6% of the entire global population died—20 to 50 million worldwide). It is the first of the two pandemics involving H1N1 influenza virus. The second was the 2009 outbreak of swine flu. Scientific studies suggest that the 1918 strain was H1N1-subtype influenza A virus. It had a higher mortality rate as compared to previous influenza epidemics.

This pandemic lasted from January 1918 to December 1920; there were

Table 5.1: Snapshot of Bioweapons

Agent	*Cause*	*Transmission*	*Systems/Organs Affected*
Anthrax	Bacillus anthracis	Inhalation, ingestion, cutaneous contact	Respiratory tract, intestinal tract, cutaneous or skin, central nervous system
Botulinum toxin	Clostridium botulinum	Open cuts and wounds are infected, ingestion of infected spores, ingestion of preformed Clostridium botulinum neurotoxin	Nervous system, muscular system
Cholera	Vibrio cholerae	Fecal-oral route, ingestion	Intestinal system
Diphtheria	Corynebacterium diphtheriae	Close contact with the discharge from an infected person's eyes, nose, throat or skin	Nose, throat, airway, heart, larynx, lungs, respiratory tract
Ebola	Filoviridae	Direct contact with the items used by infected persons or contact with the body fluids, blood and tissues of infected persons or by handling infected wild animals	Multiple *organ* systems like liver, kidneys etc.
Glanders	Burkholderia mallei	Ingestion of contaminated, bacterium has entered through a cut or wound, occupational exposure to infected animals, entered humans by coming in contact with fomites, respiratory droplets, cutaneous secretions and nasal discharge of infected animals	Spleen, lungs, liver, subcutaneous tissues and muscles
Listeriosis	Listeria monocytogenes	Transmitted through soil and water, ingestion of contaminated food	Various *organs* such as the brain, lungs and gastrointestinal tract
Plague	Yersinia pestis	A break in the skin or from a bite from infected fleas, breathing in the bacteria, which then can be transmitted from person-to-person	Lymphatic system, lungs, bloodstream
Salmonella	Salmonella	Ingestion of contaminated food or water, passing it from the feces of people or animals to other people or other animals	Cardiovascular system, nervous system, gastrointestinal system
Smallpox	Variola	Face-to-face contact, direct contact with the infected victim's body fluids or items, inhalation	Particularly the skin, but also other organs like mouth, tongue, liver and lungs
Tularemia	Francisella tularensis	Inhalation, secondary infection, through an open wound or from the bite of an infected fly or tick	Digestive system, lungs, and several other organs like spleen and kidneys

three major waves that hit the world. The second wave of the 1918 pandemic was much lethal than the first as the virus had mutated to a much lethal form. The first wave, also termed as "three-day fever," was highly contagious but caused fewer deaths. The second wave was a virulent form of the disease that spread throughout the world in 6 months. Cemeteries and mortuaries were overburdened and there was a shortfall of caskets, grave diggers and burial plots. The health systems were overwhelmed. There was no tried and tested antiviral.

There are several theories of its origin suggested by various epidemiologists and medical historians, but all these are speculations. One theory suggested that the flu originated in the Middle East and there was another theory that suggested India and China were the origin. Military speculations suggested that some of the allies thought of the pandemic as a biological warfare weapon of Germany. The best recorded first case occurred in the military outpost of Fort Riley, Kansas. This is where new recruits were trained before being sent to Europe to fight in the World War I. In March 1918, Private Albert Gitchell, a company cook, was infected and within an hour several soldiers were also infected. Similar cases were reported from various other military camps. By the end of October 1918, 195,000 Americans had succumbed to influenza. Then this flu was carried by American troops on board of troopships to Europe, and it then spread and ravaged every country that came in its path. When the flu hit Spain, Spain made public announcement of the same as it was a non-war country there was no news censorship. As most people first heard about the flu from Spanish reports, therefore, the name Spanish flu. Thereafter, the second wave struck more nations thereby creating a deadly shroud around the world. The second wave hit three port cities of Brest (France), Freetown (Sierra Leone) and Boston (United States) at nearly the same time. The third wave occurred in the winter of 1919 as a result of an armistice that brought an end to the World War I. As a result of which everyone jubilated the end of both war and flu and celebrated with each other on the streets in close contact. This celebration triggered the third wave. The third wave was not as lethal as the second wave but definitely deadlier than the first wave.

The 1918 pandemic has been described as "the greatest medical holocaust in history" as it had killed more people than any other illness in history. It is capable of infecting and killing a person on the same day. Unlike seasonal flu where most death cases are seen in the elderly and children, this pandemic caused deaths in adults in the age range of 20 to 40 years. The impact of this flu was so deep that it reduced the average life expectancy in the United States by more than 10 years. Half of the United States soldiers who died in Europe died because of influenza.

The World War I fuelled the spread of this disease. It not only swept North America and Europe, but also made its catastrophic impact felt in far flung regions like Alaska. The war, rallies, troop shipments, bond drives, etc., brought people in close contact with each other; hence the disease spread everywhere. Initially it affected military troops and camps, but later on the pandemic spread to civilian populations too.

When the 1918 flu disappeared, no samples of the virus were preserved for scientific study. However, in 1997, tissue samples from frozen victims were used to reproduce the virus for research. Scientists at the Armed Forces Institute of Pathology studied the 1918 virus and concluded that it is a unique virus but is related to the swine flu, and that this virus came into humans from pigs. According to latest research findings of 2004, researchers discovered that the 1918 virus may have infected humans directly from birds, rather than going from birds to pigs and then from pigs to humans. And this could be the reason why it was lethal as the human immune system was not prepared for fighting viruses coming directly from birds (back then).

B. As a Biological Weapon and Bioterrorism Threat

Given its short incubation period (the shorter the incubation period, the higher the risk of death), ability to cause a major health impact, availability, swiftness and pace, and its ability to resurface in a variety of virulent forms and with advances in genetic engineering to create more virulent strains, there is a possibility of using influenza virus in bioterrorism acts.

Based on past influenza pandemics and epidemiological modeling, the Wealth Health Organization (WHO) said that a future influenza pandemic would cause between 2 million and 7.4 million deaths worldwide. A study prepared for the Lowy Institute for International Policy by researchers Warwick McKibbin and Alexandra Sidorenko stated that in a "worst case scenario" more than 142 million people could be killed and there could be an economic loss of about US$4.4 trillion. Such speculations indicate the scale of loss of life and economic output if an unintentional or intentional outbreak were to occur.

Aerosol inhalation could be a possible terrorist attack scenario as influenza is easy to spread in this form. Another aspect that would make dispersion effective is that influenza virus is hard to trace, especially when it is manipulated, and this would enable the virus to spread to a wide geographic area in a short time span. Rogue states and terrorists can spread this virus by posing as international travelers and infecting co-passengers thereby spreading the virus to a wide geographic area in a short period of time.

A WORD OF CAUTION

A natural influenza attack can be an invitation to bioterrorism as the public is already in the panic mode and the health facilities are already burdened. Terrorists can use this situation to their advantage.

Moreover, influenza shares some common symptoms with other diseases like anthrax, which could cause panic and confusion, and delay in treatment as there are no point-of-care screening tests (POCT—medical testing at or near the site of patient care) to identify the exact disease when it comes to common symptoms.

C. Symptoms and Management

Symptoms of intense pain, rib-cracking coughing spells, cyanosis of the skin (bluish discoloration of the skin and mucous membranes), massive acute pulmonary hemorrhage and respiratory failure are experienced. The majority of victims who died during the 1918 pandemic succumbed to secondary bacterial pneumonia (infection of the lungs).

During the 1918 pandemic, there were no vaccines or antibiotics available. During that period, public health ordinances and measures like anti-germ masks to be worn in public places, limited time for funerals, closings of schools, churches and offices, social distancing, etc., had to be strictly followed or else heavy fines were slapped on defaulters. Public gatherings were limited to stop the spread of the disease. However, all attempts to deal with this new strain of influenza were futile.

After extensive research, drugs like Amantadine Rimantadine, Oseltamivir and Zanamivir have been developed. These drugs help to reduce the severity of the symptoms if taken soon after infection. The cause of concern is whether their current limited stock would be able to handle a potential pandemic and moreover, new strains have shown resistance to some of these drugs.

5.3.2 *Anthrax—Universally Acknowledged as a Bioweapon*

A. Introduction

Anthrax (Category A) is a disease caused by the rod-shaped gram-positive bacterium (gram-positive bacteria are those that retain the violet stain used in the Gram stain test), *Bacillus anthracis*. The name anthrax comes from *anthrakitis*, the Greek term for *anthracite*. In 1876, Robert Koch proved that anthrax was caused by this bacterium. Anthrax is most common in agricultural regions like South and Central America, Africa and Asia. Human cases of anthrax are less frequent.

The bacterium exists in nature in two forms: as an active growing cell called the vegetative form or as a dormant spore. It is capable of forming long-lived spores. Even in an inactive state, the spores can remain viable for many decades and perhaps centuries without nutrients or water.

For example, spores at animal burial sites of anthrax-killed animals have known to have re-infected animals after even 70 years of dormancy. When a spore enters a host organism, the internal environment of the host (for example, elevated CO2 and temperature levels inside an animal) causes the spore to reactivate and multiply rapidly. The spores release toxins that kill the host cells by disturbing their water balance and signaling function.

Anthrax bacterium most commonly infects domesticated and wild herbivores such as sheep, cattle and goats. It can also infect humans when they are exposed to tissues or blood of infected animals, or eat infected animals, or are exposed to spores found in animal leather, meat, wool or fur. People who are exposed to dead animals and animal products are at the highest risk. After the 2001 attack, even media personnel, government employees, mail handlers and postal officials belong to the high risk group.

The earliest reference to anthrax is found in the Fifth Plague. Even other ancient texts and writings like the Iliad, the Book of Exodus, etc., have described this disease.

There are 89 known strains of anthrax. The most widely recognized strain is the *Ames* strain, which was used in the 2001 anthrax mail attacks. It is used by the United States as a "gold standard" for development of vaccines. The *Vollum* strain is yet another virulent strain and it is much more dangerous, which was successfully weaponized during the World War II, but was never used. A variation of Vollum, *Vollum 1B*, was used during the 1960s by the United Kingdom and United States bioweapons projects. The *Sterne* strain is an attenuated vaccine strain used as a vaccine.

Inhalational anthrax occurs when people breathe in the anthrax spores, for example, occupational exposure in processes like tanning infected hides. Then these spores begin to attack the lymph nodes surrounding the lungs. As these spores begin to germinate into bacteria, swelling of the lymph nodes begins. From here the anthrax spores begin to spread to other parts of the body. Inhalational anthrax is the most lethal form. In 2006, a Scottish person died from inhalational anthrax after participating in a drumming class where contaminated African drums were used. Gastrointestinal (digestive) anthrax occurs when anthrax-infected food like infected meat is consumed. Cutaneous anthrax occurs when the spores enter the body through cuts or wounds. Anthrax is not contagious. Humans can become infected by consuming contaminated animal meats and animal products, by handling products from infected animals or by inhaling anthrax spores.

Until the 20th century, anthrax killed thousands of people and animals, but thanks to the vaccine developed by the French scientist, Louis Pasteur, and over a century of anthrax eradication and vaccination drives, the number of anthrax cases reported has gone down drastically. The disease is still present in countries where the public health facilities are not up to the mark.

B. As a Biological Weapon and Bioterrorism Threat

Perhaps the best known biological weapon is anthrax. Anthrax is hard to diagnose, especially in the aerosolized form as it cannot be smelled or seen, but it is relatively easy to manufacture and disperse. Though it does not directly spread from one animal to another or from one person to another, its spores can be disseminated through ventilation systems, objects, clothing, dead carcasses, etc., too. Because of its stability in the spore form and high mortality rate, anthrax is an attractive biological weapon.

Longevity—as compared to other biological agents, the anthrax bacterium has a long shelf life of 40 years or more. Weaponized spores can remain in the environment for long periods of time; they can remain lethal and survive for decades.

A WORD OF CAUTION

It is a possibility that the strains of anthrax used by terrorists can be genetically modified to make them antibiotic-resistant. This again is a major global concern.

During the World War I, Germany used anthrax to contaminate animal feed and livestock; animals were infected either by being fed sugar laced with anthrax or having anthrax injected directly into their bloodstream. This was done to disrupt the food supply of Allies.

Research to weaponize anthrax was performed by the United Kingdom during the World War II. The United Kingdom produced millions of linseed cakes with anthrax spores for potential use against German livestock to retaliate in case Germany used biological weapons. The field trials with anthrax bombs on Gruinard Island in Scotland left the whole island uninhabitable for almost 50 years. This goes to show how dangerous anthrax is even in experiments.

Although the Geneva Protocol prohibited biological weapons, many nations studied anthrax in the 1930s and 1940s. In 1937, Japan experimented with anthrax and in 1943, the United States began its biological weapons program, and even Canada and the United Kingdom carried out experiments and tests. Thereafter in 1972, at the Biological and Toxin Weapons Convention, more than 100 nations agreed not to produce or stockpile biological weapons. Countries like Iraq are believed to have hidden stockpiles of anthrax.

Exposure to just a few anthrax spores leads to a serious infection. Anthrax spores in an aerosol form and in a concentrated powder form have been weaponized by many nations in the past. The 2001 anthrax attacks in which letters filled with high-graded weaponized anthrax were sent to the Washington offices of senators and media outlets in New York demonstrated that concentrated spores can even be dispersed through envelopes in the mail.

C. Symptoms and Management

The virulence depends on the infection route, bacterial strains and the amount of anthrax spores the victim is exposed to.

Cutaneous anthrax is characterized by flu-like symptoms of low-grade fever, vomiting, muscle ache and fatigue. Boil-like red-brown lesions that eventually develop an ulcer with a black crust and blood-tinged drainage form on the skin, and there are swollen lymph nodes in the affected areas. In inhalational anthrax, the initial symptoms can include muscle ache, weakness, chest pain and fever, and advanced symptoms can include severe respiratory distress, shortness of breath and shock. Without immediate treatment, inhalational anthrax can cause death, and even with treatment inhalational anthrax is a deadly disease as antibiotics are effective in killing the bacteria, but they do not destroy the deadly toxins that have already been released. In gastrointestinal anthrax, initial symptoms can include acute inflammation of the intestinal tract, nausea, vomiting, loss of appetite and fever, and later symptoms are bloody diarrhea, vomiting of blood and abdominal pain.

In complicated cases like Meningeal anthrax, the anthrax bacteria enter the bloodstream. It is a complication that can arise after cutaneous, gastrointestinal or inhalational infection. Meningeal anthrax causes the infection of the central nervous system (the spinal cord and brain) and meninges (the membranes covering the spinal cord and brain). Once the bacteria enter the bloodstream, they release three lethal substances—edema factor, lethal factor and protective antigen. These substances cause bleeding, tissue decay and death of the host. Without immediate treatment and antibiotics, there are complications and the fatality rate is high. Timely treatment is needed for both stopping the growth of the bacteria and neutralizing the toxins produced by them.

People who belong to the high risk group should wear protective clothing, masks and gloves. Even mail handlers need to use this protective gear.

In mild cases, primary therapy of oral antibiotics is administered, whereas in severe cases, intravenous administration of doxycycline or ciprofloxacin plus one or two of the following is given: benzylpenicillin, chloramphenicol, cilastatin, clindamycin, rifampicin or vancomycin.

There is an effective vaccine but it is used only for certain groups of people like military personnel, people who handle imported animal hides and furs, health workers and laboratory personnel; there is no standard vaccine for the masses. Some forms of anthrax like cutaneous anthrax respond well to antibiotic treatment. Ciprofloxacin was used earlier, but now Doxycycline is the preferred antibiotic. Severe forms like inhalational or gastrointestinal anthrax need a combination therapy of intensive care, breathing support, nursing care and antibiotics. Research on an antidote for the toxins released by the anthrax bacterium is underway.

5.3.3 *Botulinum Toxin—Therapeutic yet Toxic*

A. Introduction

Botulinum toxin (also known as BTX or BoNT) (Category A) causes Botulism, which is a rare muscle-nerve paralyzing illness. BTX is a neurotoxin (nerve poison) that is secreted by the gram-positive, anaerobic (airless environment; bacteria that are capable of surviving and growing in places with little or no oxygen) bacterium, *Clostridium botulinum.*

In unfavorable conditions, this bacterium has the capability to produce an endospore—a protective structure in which it can survive in a dormant state until conditions improve. BTX destroys the muscle end-plates' capability to respond to the chemical secreted from the nerve endings. It does this by interfering with the release of the chemical messenger, Acetylcholine, thereby cutting off the signals from the brain. This blockage causes muscle paralysis.

SPOTLIGHT

Though BTX is considered as potent, yet in clinical settings, it plays a therapeutic role when used in proper amounts.

BTX is useful in the management of conditions associated with muscular and glandular over-activity like eye movement disorders, tremors, etc. It is also used for cosmetic purposes like reducing forehead wrinkles, frown lines, etc., something that we know as Botox.

Different strains of the bacterium produce different types of toxins. BTX is divided into 7 neurotoxins—A, B, C [C1, C2], D, E, F and G. Only 4 out of these 7 are known to cause human botulism. Human botulism is caused by types A, B, E and F. Botulism in animals is caused by types C and D.

The first accurate description of botulism was given in 1822 by a

German physician, Justinus Kerner. He termed it as "sausage poison." In 1944, a biochemist, Dr. Edward Schantz, isolated the botulinum toxin produced by the bacterium.

Botulism is categorized based on the mode of transmission—wound, infant and foodborne. Wound botulism occurs when open cuts and wounds are infected with the bacteria that secrete the toxin. Infant botulism occurs in infants when spores are ingested; the bacteria colonize the intestines, secret the toxin and poison the infants. Foodborne botulism occurs when humans ingest preformed Clostridium botulinum neurotoxin. Botulism through inhalation does not occur naturally, but could occur as a result of a terrorist attack or a laboratory accident (only three cases from a laboratory accident have ever been reported). Improperly stored or canned foods, infant intestinal tracts and deep wounds provide a perfect environment for the spores to develop and multiply.

B. As a Biological Weapon and Bioterrorism Threat

The Japanese terrorist cult Aum Shinrikyo tried to use BTX as a weapon through aerosol-generating equipment in the early 1990s. The cult attempted to spread it as an aerosol thrice but did not succeed.

Several countries like the Soviet Union and Iraq (Iraq produced bombs and missiles filled with this toxin) are known to have produced BTX as a biological weapon despite being signatories to the Biological and Toxin Weapons Convention. It is believed that the Soviet Union's biological weapons program attempted to splice the botulinum toxin gene into other bacteria. The United States first produced BTX during the World War II out of fear that Germany had produced it for battlefield use. Later on the United States terminated the BTX production program. The Japanese biological weapons program, Unit 731, performed human experiments by feeding *Clostridium botulinum* cultures to prisoners of war to study the lethal effects of BTX. In the post-Cold War period, some nations in the Middle East are suspected of having stockpiles of BTX.

BTX can be used as a biological weapon not only for causing illness and death, but also for burdening the medical care facilities, disrupting the food supply chain, and for causing social panic and economic loss.

Scientists have estimated that about 1 gram could potentially kill 1 million people. According to weapons experts, a small amount of BTX is capable of killing humans; this factor makes this toxin a candidate for use in weapons for bioterrorism and biological warfare. Its potency, short incubation period, limited treatability and availability make this bacterium a potential threat.

In the weaponized airborne form, the bacteria would be completely tasteless, colorless and odorless, and this is why an attack would probably go unnoticed until people exhibit symptoms. Though the toxin is moderately stable, terrorists have to be skilled in order to stabilize it when using it in the aerosol form as sunlight and dry conditions can deteriorate it. Experts believe that many of the problems related to stability and dispersal can be solved with advancements in biotechnology.

Although contaminating water reserves sounds like a feasible option, but in reality, it would be a difficult option for terrorists as large amounts of the botulinum toxin would be needed to achieve a lethal concentration, and moreover, even if large amounts are acquired, chlorine used for water treatment would destroy the toxin. Terrorists would need to develop ways that can withstand the chlorine treatment in order to effectively infect the water supplies.

C. Symptoms and Management

In outbreaks, there is no standard test to detect BTX poisoning. But it is routine to assay suspected foods, patient's feces samples and enrichment cultures for the presence of the botulinum toxin.

All types of botulism produce symptoms that affect the nervous system. Typical symptoms include dry mouth, breathing and swallowing difficulties, slurred speech and drooping eyelids, and these effects subsequently cause descending paralysis of the arms, legs, trunk and breathing muscles. In complicated cases, permanent disability or death can occur. Symptoms usually appear quickly in foodborne botulism; the onset occurs within 18 to 36 hours.

Sources of infant botulism like honey should not be fed to infants under one year old. To prevent botulism, preserved and canned foods, smoked, salted and fermented fish, and low-acid foods and low-acid juices should be properly processed and prepared. Canned food with dented or bulging lids should be disposed; the bacterium creates gases when it eats and swollen cans are a sign that the food inside might be infected.

The various botulinum toxins have varying potencies and based on these potencies, proper medical care is needed. Antitoxins and intensive respiratory care are the main line of treatment. Antitoxins are antibodies that bind to BTX to prevent it from affecting the nerve cells. They can prevent the disorder from worsening, but recovery still takes many weeks. In the case of respiratory failure, if immediate treatment like ventilator support is not administered, botulism can cause death within 24 to 72 hours. For these reasons, this deadly toxin is a Category A biological weapon. In the case of wound botulism, wounds should be treated surgically. Infant botulism is treated with immune globulin in order to facilitate early recovery and shorten the hospital stay. There is no vaccine available for the general public, but an experimental vaccine produced in the 1960s is given to laboratory personnel and military troops sent to high-risk areas.

5.3.4 Cholera—The Silent Threat

A. Introduction

Cholera (Category B) is an extremely virulent diarrheal disease caused by the ingestion of food or water contaminated with the short, curved, gram-negative bacterium (gram-negative bacteria are those that do not retain the violet stain used in Gram's method of staining), *Vibrio cholerae*. It has over 150 serotypes. Only O1 and O139 strains cause outbreaks. This bacterium was identified by the Italian anatomist, Filipo Pacini.

The main route of transmission is through the fecal-oral route. Infection is contracted when people come in contact with the vomit or stool of infected people, or when people consume water or food contaminated with this bacterium.

A WORD OF CAUTION

People with low gastric acid levels, emergency workers who handle human remains, malnourished children, international travelers and elderly people are more susceptible.

Cholera is a major epidemic disease. Since 1817, there have been seven worldwide cholera pandemics. It first emerged in the 1800s from Calcutta, India and then swept through various parts of the world causing death in its path. The first pandemic (1817-1823) started near Calcutta. Contaminated water and crowded camps caused several deaths. It spread to the Middle East, Russia, Eastern African, Central Asia, Southeast Asia and China through trade routes. The second pandemic (1829-1851) began along the Ganges River delta in India and reached the United States, Europe and Great Britain. The third pandemic (1852-1859) is considered the most deadly one. It started in Bengal and mainly affected North America, Africa, Europe and Asia. The fourth pandemic (1863-1879) began in the Bengal region and then spread to the Middle East and Mediterranean, and then it went ahead to affect North America, Europe and Africa, and the cause was returning pilgrims from the Haj at Mecca. The fifth pandemic (1881-1896) ravaged populations across South America, Africa, Asia and parts of Germany and France. The sixth pandemic (1899-1923) spread throughout Asia, Middle East, northern Africa, Russia and parts of Europe. The seventh pandemic (1961-present) began in Indonesia and spread rapidly to other countries in Asia, Europe, Africa and finally in Latin America. It was caused by a new strain called the El Tor strain. The eight and the recent pandemic began in Pohnpei.

Several ancient texts like Chinese writings and Sanskrit writings have described cholera. The ancient Greek physician, Hippocrates, had mentioned cholera in his writings.

The main reservoir of this bacterium is people, seafood like shellfish, brackish coastal waters and estuaries. This bacterium can grow and multiply in fresh and salt water. It is endemic in areas with poor hygiene and sanitation. It is common in Sub-Saharan Africa, Southern Asia and India.

B. As a Biological Weapon and Bioterrorism Threat

America's experimentation with bioweapons goes back to 1860s when the United States distributed cholera-infected blankets to American Indian tribes. During the World War I, Germany developed several agents, including cholera, for use as biological weapons. The first known use of cholera was in the World War I when Germany used it against Italy. South Africa's weapons program, Project Coast, Japanese weapons program, Unit 731 and Iraq's weapons project have studied cholera as a possible biological warfare agent. The cholera bacterium was released in the water supplies of rebel-held areas during South Africa's civil war. Japanese troops dropped cholera and typhoid cultures in ponds, wells and reservoirs in various parts of China in 1941 causing heavy casualties on both sides. North Korea is also believed to have studied cholera as a possible weapon.

Factors like the potential to overburden the medical facilities with mass casualties, a short incubation period, ease of procurement and production, and 75 percent of people infected with cholera do not exhibit any symptoms (silent dissemination) make cholera a significant threat. There is possibility that Vibrio cholerae can be used in conjunction with other bioweapons to increase its virulence.

Terrorists can use the aerosolized form to spray the cholera bacterium over crops, food supplies and water reserves. As person-to-person transmission of cholera is rare, therefore, more than the aerosol form, contaminating drinking water supplies would be more effective as it would cause economic difficulties and widespread panic. Here again, contaminating water supplies would only work when a large amount of this bacterium is used and that too in areas where there is inadequate water treatment, poor water supply infrastructure or lack of disposal of sewage. This bacterium can be genetically modified to develop more virulent strains, but again this depends on how much funds and expertise do terrorists have.

There are certain aspects like sanitation and water treatment measures like chemicals, filtration and disinfection to prevent water-borne diseases, and the insufficient expertise of terrorists to develop more virulent strains that can reduce this bacterium's effectiveness as a bioweapon. Terrorists and rogue groups will be more successful in their attack in the third world countries where cholera is endemic, and there is a lack of economic resources, clean water and sanitation provisions, and medical facilities to handle bioterrorism attacks.

C. Symptoms and Management

Incubation period is from a few hours to 5 days. About 75 percent of the people infected do not develop any symptoms. Cholera infection produces an enterotoxin (enterotoxin is a toxin in or affecting the intestines and causes diarrhea and vomiting) that causes copious, painless, watery diarrhea that can quickly lead to severe dehydration and death, if immediate treatment is not given. In severe cases, without prompt treatment, death can occur in more than 50 percent of the cases.

The most severe form is that of watery diarrhea that can lead to immediate death due to severe dehydration and kidney failure. The bacterium secretes chloride to block the small intestine's ability to absorb sodium, which then causes the intestines to create massive amounts of fluid that then produce thin, grayish brown diarrhea. The symptoms begin suddenly with intestinal cramps and diarrhea. Secondary symptoms include illness, headache and vomiting. In more severe symptoms, there is a rapid loss of body fluids, a rapid loss of more than 10 percent of bodyweight and shock—these severe symptoms can cause death within few hours.

Severe dehydration results in dry mucous membranes, renal and cardiac failure, hypokalemia (a medical condition in which there is a lower than normal amount of potassium in the blood), intravascular volume depletion and severe metabolic acidosis (a clinical disturbance in which the body produces too much acid).

In the case of cholera infection, prompt administration of oral rehydration salts (sodium and potassium) is needed to compensate for the loss of body fluids and to balance the electrolytes. In severe cases, antibiotics and administration of intravenous fluids (saline or sugar and water) are needed to reduce the volume of fluid loss, reduce the duration of diarrhea and flush out cholera through stool. Antibiotics such as doxycycline or tetracycline are used to achieve these purposes. Vaccines are used to provide antigens (an antigen is a substance that causes the immune system to produce antibodies) to bring about protective antitoxic and antibacterial immunity.

Boiled or treated water and beverages need to be consumed. Food should be consumed only after it is thoroughly cooked. Raw or undercooked seafood like fish or shellfish should not be consumed. People, especially traders and travelers going to areas where cholera outbreaks have either occurred or are likely to occur, need to strictly follow these guidelines.

A multidisciplinary approach needs to be adopted to prevent the spread of the cholera bacteria. Surveillance, preparedness, personal hygiene measures, trade and travel sanctions, safe food manufacturing and handling practices, proper water sanitation and treatment (boil water or treat it with iodine or chlorine), health education, prevention and response are needed to prevent mass outbreaks.

5.3.5 Diphtheria—The Strangling Angel

A. Introduction

Diphtheria (derived from the Greek word 'dipthera,' which means leather) is an acute infection of the upper respiratory tract and occasionally the skin caused by the gram-positive, nonmotile (not capable of movement), rod-shaped bacterium, *Corynebacterium diphtheriae*. The bacteria multiply and produce a powerful toxin that causes tissue death.

Diphtheria is characterized by a gray-brown colored membrane coating (pseudomembrane formation) in the nose, throat or airway. This disease is contracted through direct contact with nasopharyngeal secretions or skin lesions, or through respiratory droplets discharged during talking, sneezing and coughing. It is highly contagious.

There are two forms of diphtheria—cutaneous infection and respiratory tract infection. People with low antitoxin antibody levels or people who are not immunized are at the highest risk. Cases of cutaneous infections are most often seen in the tropics and also in unvaccinated travelers and in regions where hygiene conditions are poor. Outbreaks may quickly spread in areas of poverty and overcrowding.

During the 17th century, a major epidemic of diphtheria referred to as *the strangling angel of children,* swept through much of Europe. It caused the death of many children by suffocation; hence the name.

Today, diphtheria is extremely rare in developed countries due to widespread vaccination programs and surveillance and reporting systems. But resurgence of this disease is seen in developing nations due to the poor immunization coverage.

B. As a Biological Weapon and Bioterrorism Threat

A bioterrorism attack using diphtheria bacterium may not be that effective in industrialized countries as it has largely been eradicated in industrialized countries through widespread vaccination. But it could cause many deaths in countries where vaccination is not affordable or not available.

Soviet scientists spliced the diphtheria toxin gene into the plague bacterium with an aim to create new microbial weapons that would have better stability, new clinical syndromes and high virulence.

Reports also suggest that inadequate herd immunity to diphtheria can lead to outbreaks. Terrorists can use this scenario to their advantage by using the bacterium as a weapon to cause further panic and loss of life.

Diphtheria toxin, the poison secreted by the bacterium, can be used as a bioweapon too.

C. Symptoms and Management

The Schick test is a diagnostic test designed to evaluate the susceptibility to diphtheria. For diagnosis, physical examination of the mouth and tests like throat culture and Electrocardiogram are used.

Symptoms begin 2 to 5 days after exposure. The symptoms of respiratory diphtheria include fever, swallowing problem, muscle weakness, breathing difficultly, malaise, barking cough and hoarseness, and sore throat. People with a severe infection can also experience lymph node swelling and neck swelling. Further progression can lead to the obstruction in the airway with subsequent suffocation and death. Complicated cases also lead to severe conditions like myocarditis (toxic damage to heart muscles), circulatory collapse, neuritis (toxic damage to peripheral nerves) and vision problems. The symptoms of cutaneous diphtheria can include non-healing skin ulcers covered with a gray-brown membrane. These ulcers may be reddened, swollen and painful.

A WORD OF CAUTION

Without immediate treatment, complications like inflammation of the heart muscle, kidney damage, and paralysis in the neck, throat, eye or respiratory muscles can occur, and even death can occur.

The infection is treated with a combination of diphtheria antitoxin and antibiotics (such as Macrolides, Erythromycin and Penicillin); they are necessary to reduce the progression of the disease and control toxin production. Treatment in the form of intubation, tracheotomy (a surgery to open the trachea also known as windpipe), intravenous fluids, correction of airway blockages and heart monitoring is also needed in severe cases. People who have come in close contact with victims should receive immunization or booster shots and prophylactic antibiotics (antibiotics to minimize or prevent development of infection) against diphtheria. The Tdap vaccine (a vaccine that protects against diphtheria, tetanus and pertussis) is a booster immunization that offers continued protection from these diseases. There are several combination vaccines used to prevent diphtheria: DTaP, Tdap, DT and Td. A booster is needed every ten years to keep immunity high. Skin infections need to be thoroughly washed with water and soap.

The best way to prevent diphtheria is by having mass immunization drives and community surveillance programs in place, in order to achieve high coverage. People travelling to areas where there is no provision for treating this disease need to protect their health by taking booster shots and also by avoiding contact with farm animals.

5.3.6 *Ebola—Synonymous with Terror and Death*

A. Introduction

Ebola (Category A) belongs to the class of Viral Hemorrhagic Fevers (VHFs). It belongs to the Filoviridae family (filovirus). It is one of the 18 viruses that cause viral hemorrhagic fever. It causes destruction of blood vessels leading to bleeding (hemorrhaging) and swelling (edema), attack on the immunity system, and problems related to the coagulation of blood like blood clots.

Ebola gets its name from a river in the Democratic Republic of the Congo (formerly Zaire) in Africa. Ebola was first discovered in Congo in 1976. There are five identified subtypes of Ebola virus (Ebola-Zaire, Ebola-Sudan, Ebola-Ivory Coast, Ebola-Bundibugyo and Ebola-Reston); four of the five subtypes have caused this disease in humans and one has caused it in non-human primates (gorillas, chimpanzees and monkeys).

SPOTLIGHT

The natural reservoir of the ebola virus is not definitively known, but based on similar viruses, scientists and researchers believe that it is animal-borne.

The ebola virus is transmitted by direct contact with the items (clothing, contaminated needles, bed linens, etc.) used by infected persons or contact with the body fluids, blood and tissues of infected persons or by handling infected wild animals. Health care professionals have frequently been infected while treating ebola patients.

B. As a Biological Weapon and Bioterrorism Threat

Ebola has not been used as a bioweapon. Reports indicate that the ebola virus was researched and weaponized by the former Soviet Union's biological weapons program, Biopreparat. It is speculated that the Soviet Union had aerosolized the ebola virus for dissemination as a biological weapon. It is also believed that the Soviet Union developed a hybrid version of ebola and smallpox called as Ebolapox. In 1992, the Japanese cult Aum Shinrikyo made an unsuccessful attempt to obtain ebola in Zaire during the ebola outbreak there. There have been three reported incidents of researchers being infected after being pricked with contaminated syringes—the researchers from England and America recovered, but the victim from

Russia died. In 2004, while conducting research on ebola, Antonina Presnyakova, a Russian scientist, died after accidentally pricking herself with a needle laced with ebola. This accident shows how fatal and volatile ebola is even in controlled laboratory settings.

Factors like a high fatality rate, a relatively low production cost and a low infectious dose (especially in the inhalation mode—a very small number of viruses is needed to contract this infection) make ebola an effective bioweapon. Factors like lethality, stability and no effective treatment make it a viable weapon. Terrorists could obtain this virus from natural sources and spread it in the target area through person-to-person contact. Extremist groups and terrorists can send their own members (suicidal terrorists) as live ebola weapons in crowded areas or they could travel to infected areas and initiate contact with a sick person to contract the virus, and then travel to other unaffected areas. With the increasing number of suicidal terrorists, the fear of ebola virus being used as a weapon has increased. Experts believe that aerosolized ebola can be highly lethal and contagious.

According to experts, a small amount of ebola released in the subways of metro cities like Washington D.C. or New York has the potential to cause mass civilian casualties within a few days.

C. Symptoms and Management

The incubation period ranges from 2 to 21 days. Symptoms commonly include dehydration, sore throat, headache, weakness, fever, vomiting, diarrhea, and in some cases, both internal and external bleeding occur.

There is no standard treatment to assist in the early diagnosis of Ebola and there is no licensed vaccine. Therefore, preventing the spread of the virus is crucial to containing outbreaks. Oxygen to deal with breathing problems, treatment for infections, fresh or frozen plasma and antiviral agents to replace clotting factors, intravenous fluids to prevent dehydration, pressor agents to maintain blood pressure, and medications to control bleeding and to get rid of fever are needed.

In case immediate medical attention and supportive care are not given to victims, death may occur and the main reasons are multi-organ failure and massive blood loss.

Due to the highly contagious nature of Ebola, immediate identification and isolation of victims is necessary in order to break the chain of transmission. Constant monitoring of people who are in close physical contact with the victims is necessary. Doctors and health care workers need to follow proper barrier nursing precautions and infection control precautions (isolation of patients, sterilized needles and syringes, protective clothing, masks, gloves, etc.) while treating victims of Ebola. Scientists and laboratory workers need to follow biosafety level 4 (BSL-4 applies to dangerous agents that cause severe to fatal diseases and aerosol-transmitted laboratory infections, and for which there are no vaccines) laboratory precautions like self-contained breathing apparatus and biohazard protection suit.

No vaccines and drugs are available. Scientists are actively studying a possible Ebola vaccine and antiviral drug treatments. Extensive clinical trials are being carried out to come out with a viable vaccine.

5.3.7 Glanders—The First Bioweapon of the 20th Century

A. Introduction

Glanders (Category B) is an equine disease caused by the gram-negative, nonmotile, rod-shaped bacterium, *Burkholderia mallei* (formerly known as *Pseudomonas mallei*). This causative agent cannot persist in the environment outside its host. It can survive in room temperature water for up to a month and for more than a year in favorable environments (humid or wet conditions). Glanders is also known as Equinia or Farcy. It is primarily zoonotic (animal-borne). It is an infectious disease that occurs primarily in horses, but also affects mules and donkeys. It can be contracted by other animals like goats, dogs and cats too.

SPOTLIGHT

Glanders is one of the oldest diseases ever described. It was described by both Aristotle and Hippocrates. It was also described by two renowned Roman veterinarians, Apsyrtus and Vegetius. In 1664, its contagious nature was recognized. In 1821, the first account of glanders in humans was reported.

Glanders can spread to humans when contaminated meat is eaten, or when the bacterium has entered through a cut or wound, or there has been occupational exposure to infected animals (such as animal caretakers, equine butchers, laboratory workers, slaughterhouse workers or veterinarians), or it has entered humans by coming in contact with fomites (objects used by animals like grooming tools, harnesses, etc.), respiratory droplets, cutaneous secretions and nasal discharge of the infected animals. Human-to-human transmission is possible when one comes in close and prolonged contact with infected humans.

There are four forms—cutaneous, pulmonary (lungs), septicemia (bloodstream) and chronic.

Although glanders was once widespread throughout the world, it has been eradicated from many countries with the help of test and slaughter programs and disease control programs. It is still endemic in Asia, the Middle East, Africa, and Central and South America.

B. As a Biological Weapon and Bioterrorism Threat

As a biological weapon, especially in the aerosolized form, the glanders bacterium can be used to destroy livestock and humans. Because it is difficult to diagnose (there is no rapid detection assay), it is lethal, it is resistant to routine antibiotics and has a high mortality rate in humans, for all these reasons, this bacterium was one of the first agents to be used in modern warfare. Very few organisms are required to cause the disease. Despite these advantages, there is one disadvantage that can limit its use as bioweapon—it cannot persist in the environment outside its host.

Glanders was used as a weapon in both the world wars and in the American civil war as well. Its use resulted in the disruption of supply lines and artillery movement. During the World War I, German undercover agents are said to have used glanders and anthrax in an effort to destroy livestock, supply convoys and humans in the United States, Norway, Romania, Spain and Argentina. During the World War II, the Japanese developed glanders as a possible weapon against horses, civilians and prisoners of war. Several nations experimented with *glanders* in the second half of the 20th century. The Russian military forces in Afghanistan were alleged to have used glanders against opposition forces in Afghanistan between 1982 and 1984. The United States studied glanders as a possible bioweapon in 1943-44 but did not weaponize it.

Glanders was imported to the United States during the second half of the 19th century. It has been eradicated from North America, Australia and most of Europe through import restrictions, and surveillance and destruction of infected animals. In March 2000, the first reported human case in more than 50 years came to light when a 33-year old microbiologist was diagnosed with laboratory-acquired glanders; this was the first human case reported in the United States since 1945.

When human cases occur in the absence of animal cases or an epidemic, there is likelihood that such cases are a result of a biological attack. Terrorists could use this bacterium by releasing it in public places in the aerosol form or by disseminating it by direct contact with infected objects.

C. Symptoms and Management

In the late 1800s, the Mallein test (an intradermal/skin test for glanders) was developed for the diagnosis of glanders. This test along with some conventional serological tests is needed to diagnose glanders.

In the cutaneous form, swelling and sores at the site of the skin infection occur, and there can be nodules appearing along the course of the lymph vessels, particularly of the extremities. These nodules then form ulcers that release highly infectious, sticky pus. In pulmonary infection, infection in the lobes of the lungs, pneumonia, pulmonary abscesses and pleural effusion can occur. In the septicemia form, the bacteria enter the bloodstream in large enough numbers and develop symptoms like pus-filled lesions on the skin, chest pain, shivers, diarrhea, fever and respiratory distress. In the chronic form, glanders affects the skin, liver, respiratory tract, gastrointestinal tract, spleen or muscles. It is characterized by multiple ulcers, nodules and abscesses.

Once the bacteria are in the host, synthesis and release of certain toxins like lecithinase, hemolysin, pyocyanin and collagenase occurs.

The incubation period ranges from 1 to 14 days. This period and the degree of symptoms depend upon the mode of transmission, route of exposure and the site of infection. Inhalation has an incubation period of 1 to 2 days. When the agent enters through broken skin or mucous membranes (mucosal surfaces of the nose, eyes and mouth), a local infection with ulceration develops. In such cases, the incubation period is 1 to 5 days. Localized infection includes ulceration, swollen lymph nodes and increased mucus secretion or pus discharge from nose, eyes and respiratory tract. Respiratory infections, pneumonia and pulmonary abscesses can also occur. Infections in the bloodstream are fatal and have an incubation period of 7 to 10 days. In chronic cases, multiple abscesses can occur, which can affect the skin, muscles or liver. Abscesses may need to be drained. For treating these symptoms, it is always better to visit a health care provider and seek immediate medical attention. Without medical treatment, there is a 95% fatality rate.

As human cases of glanders are rare and there is no documentation about the same, there is limited information about vaccines, drugs and treatment. There is no standard recommendation or reliable treatment yet.

People working on research projects in laboratories should follow the biosafety level 3 (BSL-3 applies to infectious agents which may cause serious or potentially lethal diseases after inhalation) laboratory precautions. Laboratory workers should always wash their hands regularly and avoid touching their nose, mouth and eyes. Protection gear in the form of masks, gloves and eye protection is a must. There is no vaccine available for this disease. In the case of infected animals, the spread of the disease to other animals and humans can be prevented by eliminating the infection in the animals. Rigorous disinfection of the area (farms, hospitals, etc.) involved should be carried out. Animals and humans suspected of having glanders should be isolated.

Post-exposure prophylaxis (prophylaxis is a short-term treatment started immediately after exposure to a pathogen in order to reduce the likelihood of infection) with several potential vaccine candidates is being studied by scientists.

5.3.8 Listeriosis—Rare but Life-Threatening

A. Introduction

Listeriosis is a rare but serious type of food poisoning caused by the gram-positive, motile (capable of movement) bacterium, *Listeria monocytogenes*. It is one of the most virulent foodborne pathogens.

The bacterium is found in wild and domesticated animals, and in soil and water. Food products, fruits and vegetables can get contaminated with this bacterium when they come in contact with contaminated water, soil or manure, which in turn can cause infection in humans who consume the infected food products, fruits and vegetables. Silage is a common source of infection among farm animals.

This bacterium can be transmitted through soil and water. It can grow at low temperatures and under high salt conditions; hence high-risk foods like delicatessen meats, unpasteurized milk, pre-packed salads, paté, shellfish, meat products and soft cheeses can become contaminated despite proper storage and processing conditions. Outbreaks due to high-risk foods have been reported in the United States, Europe and Australia. Cutaneous listeriosis is rare; it is largely confined to veterinarians who handle infected animals.

B. As a Biological Weapon and Bioterrorism Threat

As it is ubiquitous in nature, as it kills at least 1 in 5 persons it infects, as it has great temperature tolerance, and there have been many instances of outbreaks of foodborne infection (like the 1985 outbreak in Los Angeles County and the 2011 United States multistate outbreak across 28 states), all these factors make the causative agent of listeriosis a potential candidate for bioterrorism.

In 2000, 29 cases of listeriosis across 10 American states were linked to the presence of Listeria monocytogenes in deli meat. In 2012, meatball-maker Buona Vita, Inc., recalled nearly 325,000 pounds of frozen, ready-to-eat meat and poultry products as Listeria contamination was discovered through routine testing by the US Department of Agriculture's Food Safety and Inspection Service and the Ohio Department of Agriculture. Although listeriosis is a rare illness, but such instances definitely indicate that terrorists can hatch bioterrorism plots by targeting areas where such outbreaks have occurred. Moreover, food contaminated with Listeria monocytogenes may not look or smell spoiled, and therefore victims would not be aware of the same.

Scientists have managed to alter the DNA of Listeria monocytogenes. Such scientific developments can be a boon for terrorists as they get more options to wreak havoc by using such bioengineered forms at points of high vulnerability in the food chain.

Listeria monocytogenes can be used by bioterrorists to contaminate the environment, cause panic and fear in the public, target food storage and processing facilities, destroy market for particular food products (for example, dairy and meat), destroy livestock and disrupt the economy. By contaminating a single food product, terrorists would not be able to achieve mass casualties, but they can use this bacterium to cause mass casualties by targeting food sources for special groups like schools, refugee camps, military bases, etc.

C. Symptoms and Management

The incubation period is between 3 to 70 days. Symptoms include headache, fever, stiff neck, vomiting, loss of appetite, nausea, convulsions, diarrhea and muscle pain. In severe cases, symptoms can also include shock, collapse and coma. In severe cases, victims can also develop blood poisoning or inflammation of the covering of the brain (meningitis). Infection in pregnant women can cause miscarriage, life-threatening infection of the newborn, premature delivery or stillbirth. In central nervous system infection, this bacterium can cause encephalitis (inflammation of the brain), meningitis (inflammation of the membranes that cover the brain and spinal cord), cranial nerve palsy and abscesses.

A WORD OF CAUTION

Particular strains of the bacterium are able to attack the heart, which leads to serious and difficult-to-treat heart infections.

Healthy people may develop few or no symptoms. Newborns, pregnant women, people with cancer, diabetes or kidney disease, elderly people and people with a weakened immune system are at a higher risk; these people should avoid soft cheeses, raw seafood, delicatessen meats, unpasteurized milk or milk products, and pre-packed foods like salad bars.

The bacterium can cause a number of serious and life-threatening clinical illnesses—pneumonia, bacteremia (infection in the blood), neonatal infection, osteomyelitis (infection in the bone), etc.

The best way to avoid listeriosis infection is by following proper food storage, handling, preparation, consumption and disposal processes. Temperature settings need to be followed too. Raw food (from animal sources like meat and poultry products) needs to be cooked properly before consuming it. All the equipments in the kitchen (countertops, knives, cutting boards and refrigerators) should be thoroughly cleaned before and after working with meats, fruits and vegetables. Fruits and vegetables need to be thoroughly washed under running tap water before use or refrigeration. Raw meat should be kept separate from vegetables, cooked foods and ready-to-eat foods. Extra care should be taken while handling unpasteurized milk and soft cheeses. One should avoid eating uncooked or undercooked ready-made meals (for example, ready-to-eat delicatessen meats). Food left at room temperature for long periods should not be consumed.

In industrial settings and food processing plants, environmental sampling and testing programs need to be carried out on a regular basis to identify and mitigate the possible risk areas.

No vaccines are available for the infection. Most people with mild symptoms require no treatment. In the case of severe infection, Ampicillin is the antibiotic of choice. Adding Gentamicin to Ampicillin for treatment brings about a synergistic result. In critical cases, hospitalization may be needed. In the case of infection in pregnant women, antibiotics should be given promptly to the victims so as to prevent infection of the newborn or fetus.

5.3.9 Plague—The Black Death

A. Introduction

Plague (Category A), also called the *Black Death* or the *Great Dying* is a zoonotic (animal-borne) disease caused by the gram-negative bacterium, *Yersinia pestis*. This bacterium is found primarily in wild rodents. Cats are also susceptible. The disease is usually transmitted by the rat flea (fleas are natural carriers of this bacterium), also known as *Xenopsylla cheopis*. It is also transmitted through direct contact with infected tissues or fluids of sick or dead animals, or by inhaling respiratory droplets, or by the bite from infected fleas.

Plague has been the cause of three of the greatest pandemics of the modern era—in the mid-6th century, in the mid-14th century (known as the Black Death that swept through Asia and Europe and lingered on for centuries) and in the early 20th century. Plague still exists in certain parts of the world.

There are three primary forms—Bubonic, Pneumonic and Septicemic. Bubonic plague infection is the most common form and it occurs due to a break in the skin or from a bite from infected fleas. Pneumonic plague is the advanced form of bubonic plague. In pneumonic plague, the lungs are affected. It is caused by breathing in the bacteria and it can be transmitted from person-to-person through repository droplets; only pneumonic plague is contagious. Untreated bubonic or pneumonic plague may result into septicemic infection. Septicemic infection can also occur by itself.

B. As a Biological Weapon and Bioterrorism Threat

Weaponized plague has always found a central place in genetic engineering, antibiotic-resistant strains, experiments, and combining plague bacterium with other agents. Recent events like two New Mexico residents infected with plague visited New York (2002), a laboratory technician from Los Angeles infected with plague, etc., have sparked renewed interest in plague

Several nations, including Germany and the United States, have explored the use of plague as a bioweapon. It has been used right from the Middle Ages to reports that indicate plague dispersion took place during the World War II—Japan dropped plague-infected fleas over Manchuria and China in 1940. The Japanese warfare program operated a plague flea factory that produced about 100 million plague-infected fleas every few days. This program also prepared for an unsuccessful attack against the United States during the last months of the World War II.

There are reports which indicate that the former Soviet Union's biological warfare program had developed a dry, antibiotic-resistant and environmentally stable form.

SPOTLIGHT

Despite the BWC, several nations like Russia and Iraq are believed to continue with covert experiments and programs focusing on weaponizing plague and other agents.

As the bacterium is easily available, both naturally and through microbe banks, is lethal even with treatment, has simple culture techniques, has a high fatality rate, has the potential for secondary transmission from person-to-person, and can be isolated and grown in quantity in a laboratory, plague is considered as a serious biological warfare agent and a bioterrorism threat. Terrorists could spread bubonic plague as such an attack is easy to initiate. A pneumonic plague outbreak would be difficult to contain as pneumonic plague is highly contagious. This attack/outbreak is likely to go unnoticed until people exhibit signs and symptoms; hence controlling the outbreak would be more difficult due to a considerable time lag between the exposure and onset of symptoms.

There have been several incidents and thefts related to the plague bacterium. In1995, a microbiologist and a white supremacist, Larry Harris, was arrested for fraudulently obtaining bubonic plague from a Maryland biomedical supply firm. In 2003, 35 vials were reportedly stolen from a laboratory at Texas Tech University in Lubbock. In 2009, 40 Al-Qaeda members died after being exposed to the plague bacterium during a biological weapons test conducted in cave hideouts in the Tizi Ouzou province in Algeria. In 2009, the frozen remains of two mice infected with the bubonic plague went missing from a New Jersey bioterror research facility.

Terrorists would need technical knowledge and skills to sustain the bacteria's stability, especially in the aerosol form, as plague once released in the air becomes sensitive to heat and sunlight. Terrorists can acquire expertise by buying it from scientists. Despite global security measures, acquiring plague samples is still possible, especially in the former USSR. A successful terrorist attack in the aerosol form would be undetectable as there would be no explosion or cloud; aerosolized plague is tasteless, colorless and odorless. A suicide terrorist infected with pneumonic plague would not look out of place in a hospital, where he can easily spread the plague. In a sudden outbreak, most likely due to a terrorist attack, the first set of cases are likely to go unnoticed as the symptoms experienced are similar to other diseases like influenza.

C. Symptoms and Management

Examination of lymph nodes, blood culture and sputum sample is done to diagnose plague.

In bubonic plague, there is a sudden onset of fever, weakness, headache, chills, nausea, vomiting and buboes (buboes is the swelling of one or more lymph nodes). Without treatment, it spreads into the bloodstream and it can develop into septicemic plague or a secondary case of pneumonic plague. The advanced stage includes symptoms like seizures, swollen lymph glands and muscular pain. Initial symptoms of pneumonic plague are chills, cough and high fever. Later symptoms of pneumonic plague include coughing up blood and pneumonia. Initial symptoms of septicemic plague are fever, chills, vomiting and nausea, and later symptoms include abdominal pain, low blood pressure and internal bleeding.

People with suspected exposure to pneumonic plague should be placed under strict observation and monitored for symptoms. Prophylactic antibiotics may be administered to protect people who have direct contact with infected patients. If symptoms are visible, antibiotic therapy should be started immediately and patients should be kept in isolation. Contacts of pneumonic plague victims should be placed under observation or given preventive antibiotic therapy. People who are traveling to a plague-infected area can be given preventive antibiotic therapy.

Plague is highly lethal if left untreated. Prevention is better than cure. Pest control of rodents and vectors, insecticides, cleanliness and sanitation prevent plague outbreaks. Infected people need to be treated in a contained-casualty setting. Measures like routine barrier protection (masks, surgical gloves, eye protection, etc.), disinfection of clothing, and surveillance and control systems should be followed strictly to limit its potential spread.

Streptomycin has been the choice of treatment; however, this drug is no longer readily available. Alternative drugs like Ciprofloxacin, Doxycycline, Chloramphenicol and Gentamicin are used. There is no standard licensed vaccine for civilian use, but clinical trials and research is going on in this area.

5.3.10 Ricin—From Potential Medicinal Use to a Bioweapon

A. Introduction

Ricin (Category B) is a potent type II *ribosome inactivating protein* toxin (stops protein synthesis by inactivating ribosomal translation) derived from the beans of the castor plant. This plant is native to Africa and found in warm climates worldwide; it has been introduced and cultivated throughout the tropical and subtropical world.

Ricin's scientific name is *Ricinus communis.* Ricin has some potential medical uses such as cancer treatment and bone marrow transplant.

Ingestion, injection and inhalation are the main modes of contracting ricin. All routes of exposure are very dangerous and can result in death. Ricin prevents the body's cells from producing proteins they need to survive and reproduce, and without the proteins the cells begin to die and this can lead to multi-organ failure. Once the cells begin to break down, the body simply collapses and this may result in death.

SPOTLIGHT

Ricin is classified by the CDC as both a chemical and a biological agent as it is a chemical toxin but has a biological source.

B. As a Biological Weapon and Bioterrorism Threat

On the positive side, the derivative products of the castor plant like lubricants, laxatives, varnish, cosmetics, etc., make life easier, whereas on the negative side, this same plant contains a lethal toxin—ricin. As little as 500 micrograms of ricin can kill an adult. Ricin is considered as an agent of both biological and chemical warfare.

As it is easily extracted as a by-product from castor beans, easily disseminated, easily available and is difficult to detect, ricin poses a potential biological weapon threat. As a biological weapon, it could be most likely used to cause poisoning through inhalation of aerosolized powder or mist. Crystalline or liquid ricin could be used to contaminate water or food supplies. In the aerosolized form, the process of lyophilization is needed; it is a process in which water is removed from a product after it is frozen and placed in a vacuum. This process is technically challenging but not impossible. Ricin can also be disseminated by inoculation via ricin-laced projectiles. Pellets of ricin or ricin dissolved in a liquid can be injected in people.

Towards the end of the World War I, the United States Chemical Warfare Service was considering the possibility of using ricin as a weapon. During the World War II, a ricin bomb, codenamed as the W Bomb, was developed by Britain, but it was never used. Even France studied this toxin, but it felt that weaponized ricin would be unsafe without an antitoxin. During the Cold War, the Soviet Union studied ricin as a possible biological weapon. During the World War II, the United States and Canada considered ricin in cluster bombs and also had plans for mass production and field trials, but these plans were never put into action. After the World War II, America's interest in ricin subsided and it began a program to weaponize sarin. Currently there is not much funding and research going on for ricin as a battlefield weapon, because countries have become more interested in other agents such as anthrax and botulinum toxin as they are more powerful than ricin.

According to weapons experts, extremists and terrorists can potentially kill large numbers of people in crowded places by releasing ricin through aerosolized means. However, ricin is not suitable for large-scale attacks as tons of ricin would be needed to cause destruction; it is more suitable for small-scale attacks. Another inhibiting factor is that ricin poisoning cannot spread from person to person.

There is documented evidence that ricin has been used for assassinations, assassination attempts and small-scale attacks. The worst, ill-famed case of a ricin attack was termed as the Bulgarian Umbrella attack. In this attack, an umbrella with a hidden pneumatic mechanism which shot out a small poisonous pellet containing ricin was used to assassinate Georgi Markov, a Bulgarian dissident. Several incidents like in 2003 an envelope containing ricin and a note threatening to contaminate water supplies was processed at a mail facility in Greenville, South Carolina, in 2004, ricin was present in the mail received at the United States Senate, and in November 2011, the United States federal authorities arrested four men in the state of Georgia on terrorism charges (these four men were accused of plotting to release ricin in Atlanta, Georgia and elsewhere throughout the United States), indicate its threat as a potential weapon of bioterrorism. Moreover, online access to knowledge about how to synthesize ricin becomes a deadly weapon in the wrong hands.

C. Symptoms and Management

The degree of symptoms depends on the route of exposure and the dose level. If the mode of poisoning is inhalation, the initial symptoms develop within 8 hours, and in the case of ingestion, the initial symptoms develop within 6 hours. When the mode of infection is inhalation, the likely symptoms are cough, weakness, fever, breathing difficulty, pain in the joints, and tightness in the chest. The next stage would be that of pulmonary edema (fluid buildup in the lungs), low blood pressure and respiratory failure, which could lead to death. When the mode of infection is ingestion, symptoms include necrosis of the spleen, liver and kidneys, abdominal pain, gastroenteritis, dehydration, stomach irritation and vomiting. The victim may experience seizures, hallucinations and bloody urine. Without proper medical treatment, multi-organ failure and death may follow shortly after ricin poisoning. In the case of injection-initiated ricin, a small amount of ricin is enough to kill the lymph nodes, muscles and tissues near the site of the injection.

In the case of ricin poisoning, decontamination of the environment is necessary. As ricin is heat sensitive, its effects can be nullified by heat above 80 degrees centigrade. Moving away from the area of exposure to an area with fresh air is a good way to reduce the possibility of death from exposure.

Currently, there is no specific inhibitor or antidote for ricin poisoning, although extensive research is being undertaken to develop and convert compounds that block the toxin into a useful drug. Supportive medical care is necessary to get the poison off or out of the victim's body. Supportive medical care includes washing affected areas (eyes, body, etc.) with soap and water or a bleach solution, intravenous fluids to deal with dehydration, administration of oxygen support, protective mask (in the case of an aerosol attack), medications to control low blood pressure, and activated charcoal to flush out the contents of the stomach. Without treatment, ricin exposure can eventually result in organ failure and death.

5.3.11 Salmonella—An Easily Accessible Biological Agent

A. Introduction

Salmonella (Category B) is one of the most common foodborne illnesses caused by the *salmonella* bacterium. It is a gram-negative, rod-shaped, facultative anaerobe in the family *Enterobacteriaceae*. It causes two main diseases—typhoid fever and gastroenteritis (inflammation of the intestines, stomach, or both.)

The Salmonella genus is divided into two different species: *Salmonella enterica* and *Salmonella bongori*. The World Health Organization categorizes Salmonella into three types—animal salmonella, nontyphoidal salmonella and typhodial salmonella. There are approximately 2000 different serotypes or subspecies of salmonella that can infect humans. These subspecies are found in different environments—soil, water, kitchen, raw meats, surfaces and poultry.

During the 1980s, there was a widespread increase in salmonella diseases, particularly in the northern United States. The most recent major outbreak of 2010 involved contaminated eggs from several United States producers that caused the recall of over 500 million eggs.

Salmonella can be contracted by ingesting contaminated food or water. Food can become contaminated by the unwashed hands of an infected food handler or can be contaminated during food processing or food handling. Other modes of transmission include passing it from the feces of people or animals to other people or other animals. Children, especially infants, are the most likely candidates to get sick. People who have an impaired immune system are also highly susceptible to infection. Salmonella bacteria can also be found in the feces of some pets, especially those with diarrhea, and coming in contact with the feces can cause infection.

B. As a Biological Weapon and Bioterrorism Threat

Japan's Unit 731, a covert biological and chemical warfare research and development unit of the Imperial Japanese Army, South Africa's Project Coast, a top-secret chemical and biological weapons project, and France, all researched the potential use of salmonella bacteria as a biological agent.

In 1984, the Unites States-based *Rajneesh* cult poisoned residents of Wasco County, Oregon, with salmonella bacteria using restaurant salad bars to influence the results of county elections, which sickened more than 700 people.

Mary Mallon, also known as Typhoid Mary and a one-woman epidemic, was an asymptomatic carrier of *Salmonella* typhi. Through her occupation as a cook, she spread illness and death in various kitchens, hospitals and households, without being sick herself ever (she refused to believe that she could make people sick, when she was not sick herself). She caused at least a dozen outbreaks of typhoid fever.

In 2001, two vials containing salmonella were sent to the then American president, but the authorities said that these vials were unrelated to the anthrax attacks.

A WORD OF CAUTION

Because of the high incidence of naturally occurring salmonella disease, a terrorist attack can go unnoticed as it can be difficult to distinguish it from natural outbreaks.

An increasingly widespread food distribution network provides opportunities for bioterrorism. Terrorists can introduce the salmonella bacterium at any stage—processing, harvesting, packing, shipping and delivery. Ecoterrorists can use this bacterium to attack the agriculture sector in order to fulfill their agenda. A terrorist-driven salmonella attack through the food supply chain would be difficult to distinguish from natural outbreaks. As salmonella bacterium is easy to grow, stock, disseminate and readily accessible, these factors make it an ideal option for terrorists to spread fear and panic. They can use the bacterium to contaminate unguarded food supplies like supermarkets or restaurants where the level of security is minimal, or to damage the agricultural economy, or to cause widespread panic, or to disrupt the food supply system. This disease can be caused with as few as 15-20 cells. Outbreaks can be caused in the target country without even entering it by contaminating imported food or fertilizers thereby making the arrest of the involved terrorists difficult.

C. *Symptoms and Management*

Symptoms of infection, which is called *salmonellosis*, typically start 12 to 72 hours after exposure. Common symptoms include vomiting, diarrhea, nausea, abdominal cramps, chills, fever and headache. The infection usually lasts 4 to 7 days. Infection is usually diagnosed by studying fecal cultures and blood samples. Countries with poor sanitation and public health vigilance have a much higher incidence of salmonellosis.

There is no vaccine available for salmonellosis. Antibiotics are not usually needed unless the infection has spread, but severely infected persons are given antibiotics like Gentamicin, Ciprofloxacin, Ampicillin, and a combination of Sulfamethoxazole and Trimethoprim. Human carriers of salmonella should undergo gallbladder removal and antibiotic therapy to get rid of the carrier state. Some strains have become antibiotics-resistant thereby making such antibiotics ineffective.

Immediate medical attention is needed to prevent the disease from spreading further to other organs like the intestines. Antibiotics are needed for cases that do not have a healthy immune system. Even infants and old people need medical care in order to prevent the disease from spreading from the intestines to the bloodstream and from bloodstream to other organs. Symptoms like diarrhea and vomiting cause dehydration; hence patients require rehydration through intravenous fluids and electrolytes. In the case of patients with a healthy immune system, the symptoms can go away without medication and treatment. A small number of victims may develop the Reiter's syndrome—it is a type of reactive arthritis that occurs as a reaction to a bacterial infection in the body. It can last for months or years, and can lead to chronic arthritis, which is difficult to treat.

People should always follow good hygiene practices and wash their hands immediately after handling pets and animals even if they are healthy, after handling poultry, milk, eggs and raw meat and when switching from preparing one type of food to another. To prevent cross-contamination of food, uncooked meats should be kept separate from cooked food. Surfaces and utensils that previously were in contact with raw meat should be thoroughly washed before reuse.

5.3.12 *Smallpox—The Terror in the Freezer*

A. Introduction

Smallpox (Category A) is an acute contagious disease. It is one of the most devastating diseases known to humanity. It is spread by Variola (it means "spotted" in Latin), which is a member of the *orthopoxvirus genus* family. It also includes viruses such as cowpox, molluscum contagiosum and monkeypox. There are two types of smallpox—variola minor, a milder form of smallpox and variola major, the most severe form. Variola major is a serious illness that can be life threatening in people who have not been vaccinated. There are four types of variola major smallpox: ordinary (the most frequent type), modified (mild and occurring in previously vaccinated people), flat (malignant) and hemorrhagic (the rarest and the most lethal form).

The earliest physical evidence of smallpox is probably the vesicular skin lesions on the mummified body of Pharaoh Ramses V of Egypt. Ancient medical writings from China and India describe a smallpox-like disease. Smallpox has been responsible for many history changing events.

Smallpox can be contracted by inhaling the virus. It is spread to others through sneezing and coughing, or by direct contact with the infected victim's body fluids or items.

B. As a Biological Weapon and Bioterrorism Threat

Smallpox epidemics plagued Asia and Europe until an English physician and scientist, Edward Jenner, developed a vaccine in 1796. It was officially eradicated in 1980 through a huge intensified program of the Wealth Health Organization, except for a few samples saved for government research. But the recent renewed interest in smallpox is from a bioterrorism threat standpoint, especially after the 2001 anthrax attacks. Reports indicate covert development is happening in several nations. According to experts, even if smallpox is introduced into a small area within a city, the virus has the potential to spread throughout the world within a year.

Most people today have no immunity to smallpox. No one knows for sure who has smallpox stockpiles. Moreover, there is limited vaccine availability in the world to deal with a natural outbreak or a deliberate biological attack. Although officially smallpox virus exists in two centers—State Research Institute of Virology and Biotechnology (Russia) and CDC (USA), there is growing global concern that there is smallpox virus stock present in secret centers. People have expressed concern that terrorists may try to get access to these stored virus samples. It is believed that terrorist groups and nations like China, Cuba, Russia, North Korea, etc., are either involved in clandestine research, or in stealing or buying this virus. Well-funded terrorist groups may bribe scientists who have the expertise to weaponize smallpox. All these factors make the bioweapon and bioterrorism threat dynamics all the more unpredictable. In current times, given these dynamics, even a single smallpox case could become a global medical crisis.

Smallpox is believed to be first used as a biological weapon during the French and Indian Wars (1754-1767) by British forces in North America. In the 1980s, the Soviet Union developed variola as an aerosol biological weapon and produced tons of virus-laden material yearly intended for intercontinental ballistic missiles.

Smallpox is potentially one of the most devastating bioweapons. Several factors like variola is environmentally stable as an aerosol, it has a high fatality rate, it is easily spreadable from person-to-person and it has no licensed treatment, make it a significant threat and an ideal candidate as a biological warfare agent. As little as 10 to 100 viral particles can cause infection. Smallpox pandemic needs only one infected person to trigger it. It can be used as a weapon in its typical way by sending an infected human in the target area or it can be dispersed through the aerosol form. In the aerosol form, smallpox is not only odorless and invisible, but it is also hard to detect.

C. Symptoms and Management

The onset of symptoms of variola major begins after an incubation period of 12-14 days. It is most contagious during the first week of the infection. Typical symptoms include high fever, chills, headache, vomiting, skin lesions, and abdominal and back pain. During the initial phase, 15% of people experience hallucinations.

In the case of aerosolized exposure, the virus multiplies in the person's respiratory tract. Once the virus enters the lungs, it begins to attack the bloodstream and then spreads to the intestines, skin, kidneys and brain.

Severe cases include extreme smallpox also called as blackpox. There are two types of blackpox: flat smallpox and hemorrhagic smallpox. In flat smallpox, the lesions do not project above the skin surface; unlike the ordinary form, these lesions never progress to pustules. Hemorrhagic smallpox is a highly fatal variety of smallpox in which hemorrhages occur in the skin and/or mucous membranes.

Smallpox is most commonly identified by the distinctive rash it causes which develops from fluid-filled bumps (pustules) to permanent, pitted scars (pockmarks). It can be confused with chickenpox, but the rash of smallpox is more painful and even its distribution is different from that of the chickenpox rash (chickenpox rash is more on the trunk than on the face and extremities). 2-3 days after the onset of symptoms, a rash of lesions begins to develop in the mouth and on the arms, legs and face. After about a week, these lesions become small blisters filled with pus. After 12 days, the blisters begin to crust over and after 3-4 weeks they become scabs that fall off.

Scientists and laboratory workers need to follow biosafety level 4 precautions. Medical personnel should handle smallpox patients with gowns, masks and gloves. Contaminated clothing, instruments and other objects should be decontaminated with water and bleach or by incineration or heat.

When used in the initial phase of the infection, the antiviral agent, Cidofovir, can be effective in treating smallpox symptoms. Victims should be isolated so as to avoid further spread of the virus. Relatives and close contacts of the patients should be kept under close surveillance and isolated if they have fever.

A WORD OF CAUTION

Routine vaccination programs have ended long time ago thereby leaving a large portion of the world population vulnerable to the smallpox virus.

As routine vaccination is no longer done due to which a large part of the world population is at risk, scientists are of the opinion that a new modern vaccine is needed to replace the old stockpile of vaccine and also education for public health community, emergency room personnel and first responders is needed for them to be familiar with smallpox and its dynamics.

5.3.13 Tularemia—A Very Dangerous Pathogen

A. Introduction

Tularemia (Category A) is an infection common in wild rodents that is passed to humans when they come in contact with body fluids or tissues of infected animals; especially hunters, farmers, campers, forest workers and trappers who come in direct contact with infected animals, particularly rabbits. It is also known as Pahvant Valley plague, Ohara's fever, Deer Fly fever and Rabbit fever. *Francisella tularensis* is the causative agent of tularemia; this bacterium is found in most places in the world. It is considered to be one of the most infectious pathogenic bacterium and for this reason the CDC identifies tularemia as a "Category A" biological weapons agent. There are two main types—Type A is more virulent and Type B is less virulent.

Arthropod vectors such as fleas and ticks can infect small animals like water rats, mice, squirrels and rabbits. This infection can be contracted by drinking infected water and contaminated food, by coming in contact with contaminated soil, by handling infected animal carcasses, by inhaling aerosolized bacteria (inhaled in a laboratory accident or while farming), or by eating meat of an infected animal.

Tularemia was first described as a plague-like disease of rodents in 1911 by the United States Public Health Service officials, Charles Chapin and George McCoy. They named the bacteria *Bacterium tularense* after the location where their investigation took place, Tulare County, California. In 1928, Dr. Edward Francis linked the causal bacteria agent of deer fly fever to the bacteria discovered by Charles Chapin and George McCoy; it was renamed as Francisella tularensis in Dr. Francis' honor. The waterborne outbreaks of Soviet Union and Europe in the 1930s and 1940s proved that it could be epidemic in nature. In the early 20th century, a Japanese physician, Hachiro Ohara, wrote of a disease that affected rabbits and those who ate rabbits. He established a link between cases of Yato-Byo in Japan (Yato-Byo is the Japanese term for Tularemia) and cases of Tularemia in the United States.

There are three main types of tularemia—inhalational tularemia caused by inhaling the bacteria, typhodial tularemia is the secondary form of tularemia that develops after a victim has contracted inhalational tularemia, and ulceroglandular tularemia which is contracted through an open wound or from the bite of an infected fly or tick (for example, a deerfly bite).

B. As a Biological Weapon and Bioterrorism Threat

F. *tularensis* is considered to be a lethal biological weapon as it has the capacity to cause illness or death, is highly infective and is easily disseminated. According to CDC, a small number of 10-50 or so organisms can cause this disease and 90-100% of the exposed population will develop symptoms. It has a 30% mortality rate if left untreated.

The causative agent is found in most places in the world and it can be easily isolated and cultivated in laboratory settings. But weaponizing the same would need considerable specialized, technical assistance to ensure its stability and effectiveness, especially from disinfectants and heat.

The bacteria could be used as a biological weapon in a number of ways.

The aerosol method is the most effective method of dispersal as this agent spreads easily once it is airborne. Release of the bacteria in an aerosolized form is the most likely method to be used in bioterrorism. A World Health Organization expert committee in 1969 estimated that 50 kilograms of this bacterium dispersed in aerosol form over a city of 5 million inhabitants would result in 250,000 incapacitating casualties and 19,000 fatalities. If a large number of people are exposed in an intentional release like a terrorist attack, then these infected people would overwhelm the health care system because tularemia is difficult to diagnose as it shares symptoms with other diseases.

Japan used tularemia as a germ warfare research weapon between 1932 and 1945. The United States weaponized the bacteria in the 1950s and 1960s, and stockpiled it until the 1970s. Practical research took place in 1954 at the Pine Bluff Arsenal, Arkansas. A mutant of F. *tularensis,* a streptomycin-resistant *strain Schu S4,* was standardized as "Agent UL" for use in the American M143 bursting spherical bomblet. Francisella tularensis in wet and dry forms (identified by the codes TT and ZZ) was examined during the "Red Cloud" tests (1966-67), which were carried out in the Tanana Valley, Alaska. The main aim of Red Cloud was to collect animal infectivity and biological decay rate data on aerosols of F. *tularensis* dispersed in a frigid field environment.

SPOTLIGHT

Since the September 2001 and subsequent anthrax attacks, concern about the potential misuse of dangerous pathogens including tularensis has increased.

C. Symptoms and Management

Symptoms depend on the mode of infection and the virulence of the bacterial strain. They begin to appear within 3 to 5 days after exposure. When tularemia enters through the skin, symptoms like extreme fatigue, headache, and painful and swollen lymph nodes and ulcers occur. Lymph nodes may become filled with pus and require drainage. In typhodial tularemia, infection of the bloodstream and organs occurs. Pneumonia may also be experienced in typhodial tularemia, especially more common in the elderly. Patients may suffer from jaundice or diarrhea. In severe cases of typhodial tularemia, multiple organs like spleen, kidney and liver can get damaged. In inhalational tularemia, chills, progressive weakness, vomiting, diarrhea, abdominal pain and abrupt onset of fever can be experienced.

People who work with animals (laboratory workers, veterinarians, trappers and hunters) should wash their hands as often as possible and wear rubber gloves while handling animals. They should also decontaminate surfaces and other areas of work with an alcohol and bleach solution.

Without medical care in terms of symptomatic and supportive care, the infection could lead to respiratory failure, peritonitis (inflammation of the thin tissue that lines the inner wall of the abdomen and covers most of the abdominal organs), chest pain, shock or meningitis (inflammation of the membranes that cover the brain and spinal cord), and eventually death. People with a poor immune system may experience a recurrence.

According to the CDC, tularemia can be treated with antibiotics like Doxycycline, Gentamicin and Streptomycin, whereas some experts believe that since it is extremely virulent, a more critical, next-generation vaccine is needed. Fluoroquinolones and Quinolones anti-infective drugs are also useful.

5.4 Pictographic Recap of Bioweapons

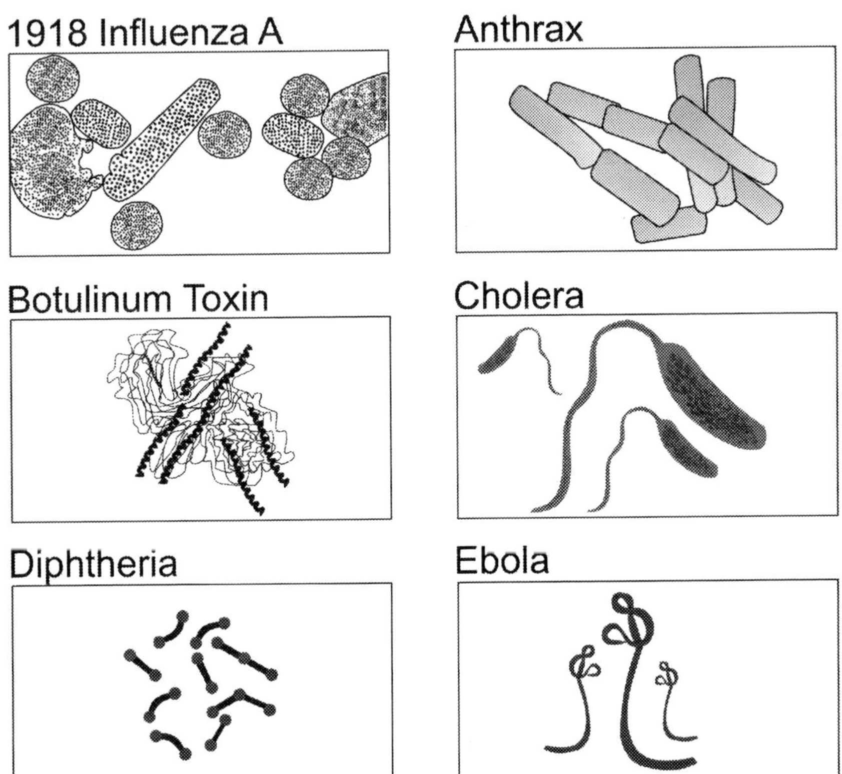

Figure 5.3: Bioweapons Part 1

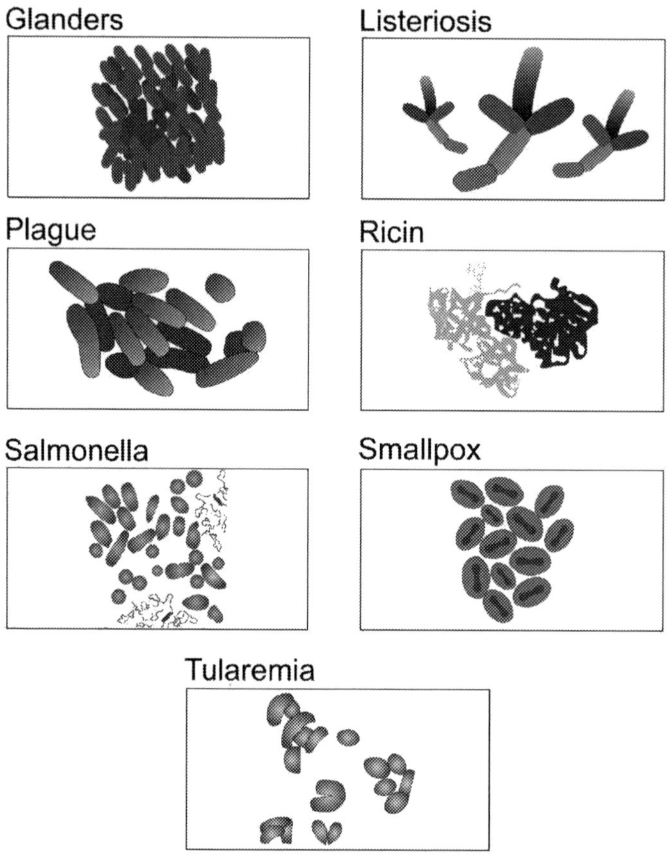

Figure 5.4: Bioweapons Part 2

5.5 Questions for Reflection

A. What are the reasons for the increasing biological weapons threat?

B. What impact does bioterrorism have on people and their lives?

C. What is being done about issues like food security, vaccines, diagnostic and detection measures, etc., in order to handle the threat of bioterrorism?

D. What policy measures are needed to strengthen the adherence to the Biological and Toxin Weapons Convention?

E. What can people do to protect themselves and their families against bioweapons?

F. How will new developments and trends in technology and science affect the implementation of the Biological and Toxin Weapons Convention?

PART V

Chemical Weapons—Bad Chemistry between Nations

6

Chemical Weapons—From Lethal Toxins to Choking Odors

Chemical weapons include weapons that are lethal and incapacitating munitions and agents. They are chemical toxins or poisons that are used intentionally to cause serious injuries, incapacitation or death during chemical terrorism or chemical warfare. A United Nations report from 1969 defines chemical warfare agents as "...chemical substances, whether gaseous, liquid or solid, which might be employed because of their direct toxic effects on man, animals and plants..." The Chemical Weapons Convention defines chemical weapons as including not only toxic chemicals but also ammunition and equipment for their dispersal.

Several chemicals used for industrial purposes are inherently dangerous and can be used as chemical warfare weapons due to one or more of the following properties: explosiveness, reactivity, toxicity or flammability.

Most modern countries can easily develop chemical weapons as there is plenty of know-how available, both in print and online form. There is a growing concern about terrorism in the 21st century as rogue states and terrorist groups with basic chemical knowledge could present a major security threat.

The common modes of exposure are ingestion, inhalation, dermal exposure, direct contact and eye exposure. Lethal doses can be attributed to accidental spills or intentional acts of terrorism. Immediate physical and medical countermeasures help to control or minimize the effects of exposure.

6.1 Delivery Systems—From Dispersion to Pyrotechnics

A variety of technologies and specialized delivery systems can be used to weaponize chemical agents, right from dispersion to pyrotechnics. For example, bombs, projectiles, spray tanks, airburst munitions, howitzer shells, etc. The main method of dispersal has been explosives in the form of central bursters, which is less efficient as compared to modern techniques, as a considerable amount of the agent is lost during incineration in the initial blast.

The effectiveness of chemical agents and the hazardous levels achieved by them depend on the delivery mechanism, method of dissemination, chemical and physical properties like irritancy, volatility, toxicity, etc., altitude of dissemination, technique of filling and storage of munitions, and accurate weather observations in terms of velocity and wind direction. The size of the particles is also very crucial. If the particles are too small, they become ineffective, and if they are too big, they fall to the ground too soon. The delivery systems should be designed in such a way that false firing due to electronic counter-measures, mutual interference and noise should be kept to the minimum.

For military purposes, chemical agents should be able to not only attack the front-line combat troops but also attack the rear-end support forces. The coverage area matters a lot. The dispersal of chemical agents should happen in such way that both the front-line and rear-end combat forces are destroyed without harming the attacker's own forces. To achieve these purposes, several delivery systems like land mines, ballistic missiles, aerial bombs, mortar shells, etc., are employed to deny terrain to enemy troops.

Dispersion is placing the chemical agent upon or adjacent to a target immediately before dissemination, so that the material is most efficiently used. Dispersion is the simplest technique of delivering an agent to its target. The most common techniques are munitions, bombs, projectiles, spray tanks and warheads. Improved dissemination techniques like thermal dissemination where pyrotechnique (pyrotechnic dissemination is effective only for non-combustible and heat-resistant agents) or explosives are used to deliver chemical agents. This technique is a major improvement over earlier dispersal techniques as it allowed significant quantities of an agent to be disseminated over a considerable distance. Thermal dissemination is the major method of disseminating chemical agents today. A more advanced technique over thermal dissemination is aerodynamic dissemination. Aerodynamic dissemination is the non-explosive delivery of chemical agents from a line source like an aircraft.

6.2 Types of Chemical Agents—Blister, Blood, Nerve and Choking

The main types of chemical weapons—Blister or Vesicant Agents, Blood Agents, Nerve Agents and Choking or Pulmonary Agents:

A. Blister or Vesicant Agents: they burn or blister the skin, mucous membranes or any other parts of the body they come in contact with. They also act as an irritant and cell poison. They are useful in reducing the fighting capability of the enemies.

B. Blood Agents: they are toxic chemical agents that affect the body by being absorbed into the blood. They inhibit the ability of the cells to utilize and transfer oxygen; they deprive the blood and organs of oxygen.

C. Nerve Agents: they are potentially lethal chemical agents that have harmful effects on the respiratory and nervous system. They interfere with the transmission of nerve impulses. They can cause death within minutes after exposure. They are the most toxic of the known chemical agents. Nerve agents have entirely dominated chemical warfare since the World War II. They were developed just before and during the World War II. In 1936, Dr. Gerhard Schrader, a German Biochemist was working on a pesticide and during his research he had discovered the effects of organophosphates on humans. These events triggered intensive research in Germany. And by the time of the World War II, Germany had N-gas, the ancestor of future nerve agents. Research continued right up till the 1940s, when the G-series was discovered. However, these efforts also triggered intensive research in the United States and Europe. Immediately after the war, research was mainly focused on having more effective forms of protection against these new agents. The end result of this research led not only to better means of protection but also to new types of agents. By the mid-1950, a set of more stable nerve agents had been developed, known as the V-series agents. The first report of these agents appeared in 1955. British chemists, Ranajit Ghosh and J.F. Newman, described Amiton, a new strain of nerve agents, as being particularly effective against mites.

The major nerve agents are the G-series agents, Sarin (GB), Cyclosarin (GF), Soman (GD) and Tabun (GA), and the V-series agents such as VE, VG, VM, VR and VX. The VX nerve agent is the most studied and documented agent, whereas the other agents in this series are less known, have not been studied extensively and information available about them is limited. The V-series nerve agents are more potent than the G-series agents. G stands for German-made (as *German* scientists first synthesized them) and V stands for Venomous or Victory. Unlike G-series nerve agents which dissipate quickly and have only short-term effects, V-series nerve agents persist for a very long time. Since the discovery of VX, there has been limited advancement in the development of new nerve agents.

D. Choking Agents: they primarily attack the lung tissue and respiratory tract. They are classified as lung damaging agents. During the World War I, both sides used them extensively, and their successful use in the battle zone set the stage for research and development of more toxic weapons.

6.2.1 *Chlorine—From a Household Name to a War Weapon*

A. Introduction

Chlorine (molecular formula: Cl) (choking agent), also known as Bertholite or Chlor, is a highly reactive, poisonous, greenish-yellow chemical, which has a choking, pungent odor. The name chlorine is derived from *chloros*, meaning green, referring to the color of the gas. Chlorine is a member of the reactive halogen group. It can become explosive and flammable when combined with other chemicals like acids or ammonia. It is a member of the halogen series. Chlorine gas may react with flammable materials as it is a strong oxidizer.

The great oxidizing potential of chlorine makes it a versatile chemical used for several purposes. Water supplies are treated with chlorine to reduce the possible spread of bacteria. It is used in industrial settings, for example, for making chlorinated solvents and resins, and it is also found in common household products like bleach, detergents and disinfectants.

The routes of exposure: inhaling the chlorine gas (for example, occupational hazard or living near a factory which uses chlorine), or ingestion of contaminated food or water, or by coming in contact with chlorine. Accidental spills (accidents at swimming pools or accidents at industrial facilities) can also cause chlorine exposure. Chlorine is one of the major causes of industrial accident toxic spills.

B. As a Chemical Weapon and Chemical Terrorism Threat

On the positive side, chlorine is used to purify metals, to make other chemicals, to make plastics, to disinfect water, etc. On the downside, chlorine is used by the armed forces and terrorist groups as a poison gas. According to the National Institute for Occupational Safety and Health, a concentration of 10 ppm (ppm stands for parts per million) is dangerous to life.

Chlorine was used as a choking agent in the early days of the World War I. Chlorine gas (also known as bertholite) was first used as a chemical weapon at Ypres, Belgium, in 1915; this attack heralded the era of modern chemical warfare. In the World War I, canisters of chlorine were opened to let the gas drift across enemy lines. Chlorine gas was used by the German army against the French, Canadian and Algerian troops at the Battle of Ypres. Initially, the attacked troops perceived the gaseous release to be a smoke screen. By the time they realized that they were being poisoned, it was too late for them to launch an attack and moreover, their rudimentary masks were ineffective against this release. All they could do was to get away as fast as possible. The troops experienced choking attacks which created a four-mile gap in the Allied line, but the Germans were not able to use it to their advantage as they too were as confused as their opponents were with this practical use of poison gas warfare of theirs. The British decided to launch a retaliatory attack to express their condemnation against Germany's use of poison gas at Ypres, but this attack turned out to be a disaster. The British used a mixture of smoke and chlorine gas, but unfavorable wind conditions caused this mixture to blow back into the British trenches. Later on gas shells were developed in order to ensure that the attacker's troops are not harmed from a reversal attack and the range of attack is extended.

SPOTLIGHT

After the World War I, chlorine by itself was not considered a major player on the battlefield, but it played a crucial part in the process of production for more lethal compounds to follow.

As chlorine is detectable due to its odor and color, and the poisonous fumes can be avoided with measures like a wet cloth or gas mask, makes it a less desirable choice as a chemical weapon. More lethal chemical agents like mustard gas and phosgene are preferred to chlorine.

Chlorine is not only a source of unintentional release from industrial accidents, but it also poses a major threat from intentional release by terrorists. As chlorine is needed for industrial purposes on a large scale, huge quantities of chlorine are processed, stored and transported, and most of the times transportation and storage is done with little or no security and this could make chlorine an attractive weapon for terrorist groups to use in highly populated areas without having to manufacture or transport it themselves. Trucks, railcars and containers carrying chlorine represent serious security vulnerability—by just puncturing or putting a hole in them, terrorists can release a lethal cloud of chlorine without being caught by intelligence agencies. These dynamics call for immediate remediation of security loopholes in the processing, storage and transportation of chlorine, and also a close tab needs to be kept on the retailers, distributors and manufacturers of chlorine.

Recent attacks involving chlorine like the insurgents using chlorine bombs against the local population and coalition forces in the Iraq War raised some serious questions regarding safety standards of chemical storage facilities and infrastructure. Insurgent groups in Iraq have been using bombs combining explosives with chlorine, something called as Improvised Explosive Devices, against Iraqi police and civilians. According to the United

Nations Monitoring, Inspection and Verification Commission (UNMOVIC carries out ongoing monitoring and verification to check Iraq's compliance with its obligations not to reacquire the same weapons prohibited to it by the Security Council.), at least 10 attacks involving chlorine have occurred in Iraq up to June 1, 2007. This is a new tactic used by insurgent groups, which adds another deadly dimension to terrorism as chlorine may not be as potent as nerve agents such as sarin, but it is a doable weapon as it does not need a complex delivery system or a complex chemical process like synthesis to cause damage and casualty.

C. Symptoms and Management

Skin, lungs and teeth examination help to determine the effects of chlorine exposure.

The symptoms depend on the amount of chlorine, length of exposure and the mode of exposure. Other factors like weather conditions, bad habits like smoking, heredity and previous exposure also determine the severity of symptoms and health effects. People with heart or lung diseases and smokers are more at risk.

Short-term, high-level exposure can cause vomiting, convulsions, coughing, blistered skin, breathing problem, burning sensation in the eyes, nose, ears, lips or tongue, blurred vision, chest pain or excess fluid in the lungs. A more severe exposure level can cause complications like skin tissue burns, bronchitis, pneumonia, collapse, and even death. Chlorine is harmful to organisms living in soil and water.

Moving away from the area of exposure to an area with fresh air is a good way of limiting the exposure. Clothing and other items like keys, wallets, etc., should be sealed within a plastic bag which should be placed in another plastic bag. Contaminated clothing, objects and victims should be handled with the help of gloves. Affected areas of the body, skin and hair should be washed with water and soap. Eyes should be thoroughly cleansed with water. As chlorine is heavier than air, it tends to settle down in low-lying areas, due to which the best way to avoid exposure is to move to higher ground. Special care needs to be given to exposed children because they are at higher risk. People with preexisting lung or respiratory illnesses should be kept under observation for at least 24 hours in hospital settings. For limiting occupational exposure, safe work practices like protective gear, proper handling and storage, engineering controls, medical monitoring program and emergency measures need to be followed.

There is no standard antidote. Medical measures like surgical removal (in the case of skin damage), activated charcoal, masks, oral drugs and antihistamine ointment, and supportive and symptomatic care help in controlling the symptoms.

6.2.2 *Chloropicrin—The Fumes of Choking Fear*

A. Introduction

Chloropicrin (molecular formula: CCl_3NO_2) (choking agent), also known as PS, Acquinite, Trichloronitromethane or Klop, is a highly toxic, colorless to light green, heavy, oily liquid with a very sharp, irritating odor. It is a strong irritating tear gas (lachrymatory agent). It exists as a liquid at room temperature. Chloropicrin is more toxic than chlorine. It was discovered in 1848 by a Scottish chemist, John Stenhouse. When chloropicrin comes in contact with strong oxidizers, combustible liquids or combustible solids, it can cause explosions. Chlorination of water can lead to the formation of chloropicrin.

Chloropicrin is used for treating timber and wood. It is used in agriculture, especially in high-value crops like tomatoes, strawberries, etc., as a pre-plant soil fumigant to control pests, fungi, nematodes, bacteria and other harmful organisms.

Exposure can occur through ingestion, inhalation, or direct contact with the skin or eyes. Exposure can also occur in residential areas that are close to agricultural areas due to an off-site drift of chloropicrin.

B. As a Chemical Weapon and Chemical Terrorism Threat

On the one hand, in industrial usage, chloropicrin is used as warning agent for evacuating people from the structures/premises to carry out fumigation, whereas on the other hand, it has been also used extensively in military operations.

Because of its high stability (it is not easily broken down), acute toxicity, easy absorption at any temperature, and as a threat of a significant explosion hazard, especially in situations that involve fire, friction and heat, chloropicrin is a strong candidate for chemical warfare and terrorism. When chloropicrin is mixed with other tear gas agents like CN or CS, a very toxic weapon is created.

A WORD OF CAUTION

Chloropicrin is fatal by all routes of exposure, especially the inhalation route.

Chloropicrin was used in large quantities as a poison gas during the World War I because of its potent activity as a lachrymator (a substance that causes tears, pain and even blindness). Russia was the first major nation to use chloropicrin. Britain used chloropicrin with chlorine in a mixture called the Yellow Star gas. The Germans used chloropicrin that caused the allied soldiers of the Italian front to experience tears, vomiting and coughing, and this in turn, made them remove their masks exposing them to the simultaneous attack of other toxic agents. It was used to fill artillery shells by the French troops. Although chloropicrin is not as dangerous as phosgene, it did manage to penetrate most of the early gas masks until activated charcoal was introduced to absorb the chloropicrin vapors. During the World War II, chloropicrin was stockpiled. The former Soviet Union had combinations of chloropicrin with phosgene and mustard agent in their chemical weapons stockpile. However, after these wars, chloropicrin's importance as a chemical weapon has diminished because of the development of more lethal weapons like nerve agents. It is no longer a weapon in the war zone, and its former application as a training agent, casualty agent and riot-control is now rare. Now, it is widely used for industrial purposes like fumigation, organic synthesis, etc.

While chloropicrin would probably not cause a huge number of deaths in a terrorist attack, it would still be successful in burdening the public health system and result in costly cleanup operations. In 1995, an unidentified man poured chloropicrin from a bottle outside a restaurant in Maebashi, northwest of Tokyo, which affected several people. Chloropicrin is capable of causing temporary blindness, which in turn, may lead to accidents. Terrorists can take advantage of such accidents to worsen the situation. For terrorist groups, procuring or producing chloropicrin and creating a improvisational weapon out of it is easy, but it requires some special impermeable protective equipment like gas mask, protective clothing, clean air respirator, etc., to effectively weaponize chloropicrin, as this chemical is known to penetrate most masks.

C. *Symptoms and Management*

The health effects and symptoms depend on the concentration and duration of exposure. Chloropicrin is severely irritating to the eyes, skin and lungs. It also interferes with the oxygen transport to the tissues.

When ingested, chloropicrin can cause diarrhea, vomiting, and burns to the mouth, throat and esophagus. Severe exposure can be fatal. In the case of inhalation, symptoms like dizziness, tightness in the chest, breathing problem, sore throat and vomiting are experienced, and in severe cases, pulmonary edema and even death can occur. Severe and prolonged skin exposure can cause corrosion and dermatitis. Severe and prolonged eye exposure can result in corneal edema and blindness. Chloropicrin can cause the throat and stomach muscles to rip because of the force of vomiting.

Victims need to be taken away from the site of exposure. In severe cases of exposure, administration of humidified, oxygen-rich air may be needed. Emergency procedures like protective gear and equipment (chemical resistant suit and boots, self-contained breathing apparatus, masks, gloves, aprons, etc.) should be followed by medical personnel and rescue workers to protect themselves during rescue operations. The affected areas (eyes, body, etc.) need to be flushed thoroughly with water for at least 15 minutes and the chemical needs to be carefully removed from clothing, shoes and other objects. Ventilation can be done to reduce the concentration level. Do not induce vomiting or administer diluents to victims who are either unable to swallow or are unconscious, unless told to do so by a medical expert. Unconscious victims should be kept on their sides to avoid choking.

Emergency medical measures like poison control, continuous observation, artificial respiration, supplemental humidified oxygen and muscle relaxation agents need to be followed in order to avoid further progression of symptoms.

6.2.3 *Cyanide—A Deadly and Quick-Acting Chemical*

A. Introduction

Cyanide (molecular formula: CN-) (blood agent and a tissue toxin) is a potentially deadly and fast-acting chemical with a bitter almond odor. The term cyanide has always been an enigmatic source of fear for humans. Cyanide is a fatal chemical that affects the cells' ability to use life-sustaining oxygen absorbed from the bloodstream; hence the cells cannot produce energy due to lack of oxygen resulting in degeneration of cells and eventually the victim's death.

Cyanide takes various forms—a colorless gas, such as Cyanogen Chloride (CNCl) or Hydrogen Cyanide (HCN), or a crystal form such as Potassium Cyanide (KCN) or Sodium Cyanide (NaCN).

Cyanide can be produced or extracted from many everyday products like silk, plastics, etc. It is found in industrial emissions. It is also found naturally in some foods and plants such as tapioca, lima beans, cherry pits, almonds, cassava, etc. Cyanide gas is used to exterminate pests. Cyanide salts are used for metal cleaning and electroplating. Cyanide is used in manufacturing textiles and paper. The most commonly used form of cyanide, Hydrogen Cyanide, is used for making synthetic fibers and resins. Certain chemicals when ingested can be converted by the body into cyanide, for example, such chemicals are found in solvents and artificial nail polish removers. Certain grasses such as Johnson grass can cause poisoning in grazing animals.

Cyanide exposure can occur as a result of ingestion (consuming contaminated water or food), inhalation (breathing air containing cyanide dust, industrial emissions, and other airborne sources like cigarette smoke or cyanide fumes from burning of plastic and rubber) and by coming in contact with the skin or eyes. Smoking cigarettes and breathing second-hand smoke also exposes people to very small amounts of cyanide. People who handle contaminated soil can also be exposed when they touch their mouth or eat with dirty hands. Accidental exposure is another route, for example, the 2000 Baia Mare cyanide spill—a major spill of *cyanide* was released into the river system that caused widespread damage in three countries. This spill has been called the worst environmental disaster in Europe since the Chernobyl accident.

A WORD OF CAUTION

Chronic consumption of cyanide-containing foods can lead to poisoning.

B. As a Chemical Weapon and Chemical Terrorism Threat

Cyanide and its derivatives have been used extensively in both military and non-military settings. They have been used in homicides, suicides and executions.

Cyanide has been used as a weapon since the days of ancient Rome. The Roman Emperor Nero's preferred poison is said to have been cyanide-containing cherry laurel water. During the Franco-Prussian War (1870–1871), Napoleon III asked his troops to increase their fighting efficiency by dipping their bayonets in cyanide.

France and Britain used cyanide bombs in the World War I, but without much success because of its high volatility and ineffective delivery mechanism. Japan and United States maintained cyanide stockpile during the World War II. Zyklon B, a cyanide-based pesticide (a form of hydrogen cyanide) invented in the early 1920s, was used by the Germans as an agent of mass annihilation against millions of enemy soldiers and civilians in Nazi gas chambers in the World War II. Cyanide was used in the United States to execute prisoners in the gas chamber from 1924 to 1999. Iraq has always shown a particular interest in chemical weapons. Iraq is believed to have used cyanide (hydrogen cyanide gas) along with other chemical agents against the Kurds during the Iran-Iraq War in the 1980s. A combination bomb of hydrogen cyanide, nerve gas and mustard gas was perhaps the worst weapon used against the Kurds in 1985. Cyanide gas precursor compounds (potassium cyanide and dilute sulfuric acid) were found in several subway restrooms in Tokyo following the sarin attack in Tokyo in 1995. These recent events show that cyanide continues to pose a military and terrorist threat.

As there are several sources of cyanide available, terrorists can easily acquire the same, but for them to make cyanide an effective weapon, especially for outdoor release, they will need expertise to achieve sufficiently high concentrations. Terrorists may find cyanide more suitable for attacks in enclosed spaces. They can also use it to contaminate food or water supplies. Weaponized cyanide can be delivered by munitions from canisters, bombs or artillery. The effective way to achieve high concentration is through large munitions. A variety of means of delivery and ready availability increases the potential of terrorist strikes. Terrorists were plotting a cyanide poison attack on London Olympics 2012 as evident from detailed instructions posted on a certain website. Such recent terror plots surely raise the current global terror threat level.

C. Symptoms and Management

The health effects and symptoms depend on the form of cyanide, route, length and concentration of exposure, heredity, and personal habits like smoking. Health care and laboratory personnel, smokers and people who work in industries (for example, metal cleaning, electroplating, plastics, etc.) where cyanide is used belong to the high-risk group. Inhalation of the gas causes the most rapid onset of toxicity.

A small dose of cyanide can cause dizziness, weakness, nausea, vomiting, rapid breathing and increased heart rate. Prolonged high-level exposure can cause low blood pressure, seizures, loss of consciousness, respiratory failure and even death. Prolonged low-level exposure can cause nerve damage, which in turn can affect muscle coordination, vision and hearing. The thyroid gland can also be affected resulting in slowing down of the development process in children and affecting the metabolism process in adults. Skin exposure can cause irritation and sores. Eye exposure can cause burning and irritation. Cyanide exposure is most harmful to the brain and heart which rely heavily on oxygen.

The victims should be moved to fresh air. In the case of cyanide gas exposure, as cyanide gas is lighter than air it will rise up, therefore, the best way to avoid or minimize exposure is by staying in the low grounds. The affected skin areas need to be thoroughly rinsed. Contaminated clothing and other objects should be removed in order to get rid of leftover cyanide; contaminated objects (clothes, wallets, watches, purses, etc.) should be placed in a plastic bag and then this first plastic bag should be sealed in a second plastic bag. In the case of cyanide ingestion, the victims may need to undergo stomach pumping. In severe cases, observation and treatment in hospital settings, care from poison control center, intravenous fluids, standard decontaminates, airway support and management, antidotes like sodium thiosulfate and sodium nitrite, high flow oxygen supplementation, drugs for medical conditions like seizures, and supportive medical care are needed to reduce further poisoning and lessen the effects of exposure.

6.2.4 Cyanogen Chloride—A Fast Killer

A. Introduction

Cyanogen Chloride (molecular formula: *CNCl*) (blood agent), also known as CK or Chlorocyanogen, is a poisonous, colorless gas or liquid with a very pungent "biting" odor. It is a highly toxic chemical that interferes with the body's ability to use oxygen. It is a clear liquid at very low temperature and a colorless gas at room temperature. CK is slightly soluble in water, but dissolves readily in organic solvents. As CK is highly unstable and reactive, it can explode unexpectedly.

CK is used in several manufacturing operations to make a variety of products. It is used in industrial settings like organic synthesis, metal cleaning, electroplating, mining, dyeing, etc.

Exposure can occur by breathing in cyanogen chloride vapor, or by consuming contaminated food or water, or by skin contact or eye contact.

B. As a Chemical Weapon and Chemical Terrorism Threat

Cyanogen chloride can be dispersed through a liquid spray (liquid spray can be used to contaminate water, food or crops) or through aerosolized means. At low concentrations, cyanogen chloride is more effective as a weapon than hydrogen cyanide as it is less volatile than hydrogen cyanide.

As CK is highly toxic, unexpectedly reactive and is capable of hampering the body's mechanism of using oxygen and eventually a number of vital organs and systems, including the brain, respiratory system and central nervous system thereby leaving the victims defenseless, it acts as a fast killer and this chemical property makes CK a potential chemical warfare agent.

According to experts, cyanogen chloride is capable of penetrating the filters in gas masks, which makes it a potential chemical warfare weapon. CK may also react with the materials in gas mask filters thereby reducing the effectiveness of filters.

It is suitable for trench warfare; CK is more dense and less volatile than hydrogen cyanide, and therefore less likely to be blown away by the wind and more likely to settle down into trenches.

CK, along with hydrogen cyanide, was first used in large quantities by the British and French during the World War I. The United States evaluated CK as a potential weapon but rejected the idea. During the World War II, United States reexamined the value of CK as a weapon, and it did maintain CK bombs, but they were not used during the World War II. After the World War II, CK was replaced by more deadly weapons.

In current times, though the production of CK for military purpose may not be going on in full swing, but CK is still produced for industrial purposes, and this fact makes CK a potential chemical terrorism agent. As CK is effective in closed spaces, terrorists could use CK in trains, buildings, malls, etc., where it is highly effective in disrupting public life and creating a sense of terror and fear in civilians.

C. Symptoms and Management

Cyanogen chloride is broken down in the body to release cyanide. CK has systemic effects, particularly affecting those organs most sensitive to low oxygen levels. Typical symptoms can include nausea, vomiting, edema, confusion, tightness in the chest, coughing, choking, sore throat, loss of consciousness and paralysis. At low concentrations, cyanogen chloride can irritate the lungs, mucous membrane and eyes, whereas at high concentrations, it can lead to paralysis and even death.

A WORD OF CAUTION

Cyanogen chloride affects organs that are most sensitive to low oxygen levels, for example, the blood vessels, lungs, brain and heart.

Depending on the concentration level, type of effects and exposure, treatment in the form of oxygen support, treatment of skin burns, chemical-resistant gloves, suit and boots, decontamination, antidotes, masks and face shields, cool compresses and continuous cardiac monitoring are needed. As pulmonary edema may develop relatively late after exposure, it is necessary to keep the victims under close observation.

6.2.5 *Diphosgene—A Valuable Reagent Plus a Poison Gas*

A. Introduction

Diphosgene (molecular formula: $C_2Cl_4O_2$) (choking agent), also known as DP or Trichloromethyl chloroformate, is a moderately persistent, colorless compound that smells like new-mown hay, grain or green corn. DP has two forms—a colorless gas and an oily, colorless liquid. It is not only a respiratory irritant but also a lachrymator. DP is prepared by radical chlorination of methyl formate or methyl chloroformate under UV light. DP is related to phosgene. Its vapors have the odor of phosgene.

At room temperatures, DP is a liquid, but rapidly vaporizes into a colorless gas. DP degrades easily into chloroform and phosgene. When the temperature is raised to 300 degrees Celsius, DP decomposes to phosgene. DP reacts violently with steam, moisture or water.

As DP is a safer substitute for phosgene and convenient to handle than phosgene, it has replaced phosgene in some laboratory and industrial uses;

DP is therefore useful for reactions traditionally relying on phosgene. It has replaced phosgene in some large-scale industrial reactions such as the formation of isocyanates and carbonates.

B. As a Chemical Weapon and Chemical Terrorism Threat

The toxicity of DP is the same as phosgene as it decomposes to phosgene on heating. The lachrymatory effect makes DP more easily detectable than phosgene. When released in high amounts, DP has the capacity to suffocate the enemy.

DP was developed during the World War I; it was a product of the chemical weapons race during this period. DP was originally developed as a pulmonary agent, a few months after the first use of phosgene. DP was created by combining phosgene with chloroform, which had the potential to destroy the filters in gas masks as the standard gas masks of those times were designed to filter out phosgene only. It was used as a poison gas in German-made artillery shells. It was first used on the field at Verdun in 1916 probably as a response to the French use of phosgene that same year. DP (codename: Surpalite) was later used by France during the World War I.

As DP is commercially available, it becomes a likely threat as a terrorist weapon. In current times, nerve and blistering agents have replaced choking agents, including DP, as the leading chemical warfare weapons. This World War I weapon is now mainly used in organic synthesis.

C. Symptoms and Management

The symptoms and toxicity depend on the concentration and the length of exposure. The onset of symptoms for low exposure begins after 3 hours, whereas high exposure results in immediate symptoms. Because of the low water solubility of DP, victims often inhale large amounts of vapor before symptoms appear. As DP is heavier than air, it tends to accumulate in low-lying areas for a prolonged period due to which children are at a higher risk.

Inhalation can cause hemorrhage, tightness in the chest and painful breathing. Ingestion can cause damage to the digestive tract, chemical pneumonitis (inflammation of the lungs caused by inhaling toxic fumes) and pulmonary edema. Eye exposure can cause burns. Direct eye exposure to the liquid form can cause ulcers and abrasions. Dermal exposure can cause skin burns. Dyspnea (shortness of breath), frothy sputum, low blood pressure, pulmonary edema, respiratory failure and even death can occur at higher concentrations.

Moving away from the hot zone into an area of fresh air is the best way to avoid or minimize exposure. Contact with the liquid form needs decontamination and burn therapy. In severe cases of inhalational exposure, artificial respiration and supportive care are needed to deal with airway obstruction and injury in order to lessen the chances of pulmonary edema. Early use of a positive pressure mask may be beneficial for preventing or treating pulmonary edema. For dermal and eye contact exposure, copious irrigation is important. For low blood pressure victims, standard resuscitation with first-line agents (crystalloid fluids) and second-line agents (vasopressors: medications to raise reduced blood pressure) needs to be performed.

6.2.6 GV—A Part of a New Series of Nerve Agents

The classical G-series agents include agents like tabun, sarin and soman. There are other variants like cyclosarin. The classical V-series agents include agents like V-gas (Russian equivalent of VX) and VX. And in between these two series is the intermediate volatility and intermediate lethality agent, **GV**. GV (molecular formula: $C_6H_{16}FN_2O_2P$) (nerve agent) is a part of a new series of nerve agents with properties similar to both the G-series and V-series. It has volatility between the G-series and V-series agents. The volatility of GV compounds is between VX and sarin due to which they are effective when penetrating uniforms. GV compounds are typically used for a semi-persistent to persistent effect. GV is a highly poisonous vapor and a potent acetylcholinesterase inhibitor (acetylcholinesterase inhibitor is a chemical that disturbs the function/blocks the action of the acetylcholinesterase enzyme, which hydrolyzes the neurotransmitter acetylcholine). Drugs such as atropine, benactyzine and HI-6 are used for treating GV exposure. There is limited information available about these compounds.

6.2.7 Hydrogen Chloride—Corrosive to the Core

A. Introduction

Hydrogen Chloride (molecular formula: HCl), also known as Anhydrous (water-free) gas or Aqueous hydrogen chloride (i.e., Hydrochloric acid), is a colorless-to-slightly-yellow corrosive, non-flammable, fuming gas with an irritating, sharp odor (at room temperature). This odor is an early warning signal. When hydrogen chloride comes in contact with atmospheric humidity, this gas forms dense white vapors of *hydrochloric acid* (hydrochloric acid is a solution of hydrogen chloride in water). HCl readily dissolves in benzene, ether and alcohol. When HCl comes in contact with metals, it produces hydrogen gas, which has the potential to cause explosion. Significant release of HCl occurs when certain plastics such as PVC, coal or waste are burned. HCl is released naturally from volcanoes.

Most hydrogen chloride is used in the production of hydrochloric acid. It is used in both commercial and household settings. It is used in making water treatment chemicals, production of vinyl and alkyl chlorides, delinting cotton, etching concrete, electroplating, refining mineral ores, manufacturing dyes and fertilizers, food processing, metal pickling, house cleaners, pigments, leather tanning, etc.

Exposure can occur through inhalation, skin or eye contact with vapors or aqueous solution, or ingestion. Occupational exposure and accidental spills is another route of exposure. Children and people working in industries where HCl is used are at a higher risk.

B. As a Chemical Weapon and Chemical Terrorism Threat

Hydrogen chloride was used as a chemical warfare weapon in the World War I, but was not an effective agent as it was not persistent.

As hydrogen chloride is corrosive and reactive in nature, terrorists will need to have special expertise in handling the same. Secret laboratories often generate hydrogen chloride gas by combining sulfuric acid with sodium chloride. Railcar delivery of anhydrous hydrogen chloride is a very risky issue as it can either be the cause of a serious accident or it can be the focus of terrorist attacks targeting railcars as chemical weapons in populated locations. Moreover, with limited emergency response and security measures on the railcar transportation network, terrorists have more scope to carry out their activities.

C. Symptoms and Management

Hydrogen chloride forms hydrochloric acid (aqueous solution) on contact with water found in the body tissues, which can be *corrosive* to the eyes, skin and mucous membranes. High concentration is irritating and corrosive to the eyes, nose, respiratory tract, skin and mucous membranes. When one comes in contact with the liquid form of hydrogen chloride, it can cause severe burns, frostbite (destruction of the skin and underlying tissue caused by exposure to extreme cold) and deep ulcerations. When inhaled in low concentrations, hydrogen chloride can cause shortness of breath, cough, and irritation in the throat, lungs and nose, and in the case of severe exposure, it can cause pulmonary edema. Mild eye exposure can lead to a burning sensation, redness and swelling, whereas severe eye exposure can cause glaucoma (an eye disorder in which the optic nerve is damaged and this could lead to permanent vision loss) and cataract. In the case of mild skin exposure, irritation, inflammation and redness are experienced, whereas in severe cases, deep chemical burns can occur. In ingestion-induced exposure, severe injuries to stomach, esophagus, lips and mouth can occur. Prolonged exposure can cause skin rashes, discoloration, etching and erosion of teeth, and chronic bronchial irritation.

A WORD OF CAUTION

Children are more sensitive to hydrogen chloride due to their greater lung surface area-to-body weight ratios and increased minute volumes-to-weight ratios. People with skin, respiratory or eye conditions may be at a higher risk too.

There is no antidote for hydrogen chloride poisoning. Measures like immediate medical attention, symptomatic and supportive care, moving to an area with fresh air, turning off furnaces, heating or air conditioning systems and fans, removal of contaminated clothing and other objects to avoid secondary contamination, and flushing the affected areas with water are needed to control the symptoms. In rescue operations, the medical professionals and rescue personnel should be equipped with self-contained breathing apparatus and chemical-protective clothing.

6.2.8 Kolokol-1—The Knockout Gas

A. Introduction

Kolokol-1 (incapacitating agent), also known as psycho-chemical gas, is an extremely dangerous opiate-derived incapacitating agent. Though the exact chemical nature is not known, kolokol-1 is believed to be a derivative of the potent drug fentanyl. Kolokol-1 is used as a crowd control or riot control agent. It is a synthetic opioid. Opioid is a chemical that has opium-like effects. It is a substance that affects the opiate receptors on the surface of the cells.

B. As a Chemical Weapon and Chemical Terrorism Threat

According to Lev Fyodorov, a former Soviet chemical weapons scientist who now heads the independent Council for Chemical Security in Moscow, kolokol-1 was originally developed in a secret KGB laboratory in Leningrad during the 1970s. And in the subsequent years, methods of dispersal were developed and tested by releasing harmless bacteria through the subway system ventilation shafts first in Moscow and then in Novosibirsk.

It is believed that the Spetsnatz, the Russian Special Forces, used kolokol-1 in Moscow in 2002 to end the Moscow theatre hostage crisis. It is believed that the chemical agent used against the Chechen terrorists in this incident was kolokol-1. This incident was a very unfortunate one as even the hostages died during this operation due to kolokol-1, which was dispersed as an aerosolized incapacitating agent.

Over the past several years, civilians too have been exposed to several chemical warfare agents, including kolokol-1. On the one hand, kolokol-1 can be used for law enforcement situations, but on the other hand, when it falls in the wrong hands like terrorists, it can be used against defenseless civilians too.

C. Symptoms and Management

The effects and degree of symptoms depend on the length of exposure, level of opioid tolerance and amount of exposure. Symptoms can include immobilization (physical or mental impairment), lethargy and reduced reaction time. In severe cases, exposure can lead to life-threatening conditions like hypoventilation (breathing at an abnormally slow rate), apnea (temporary absence of breathing) and airway obstruction. Exposure through inhalation causes rapid unconsciousness within one to three seconds and the victims remain unconscious for several hours.

In severe cases of apnea, immediate ventilator assistance should be administered. Apnea and airway obstruction can be reversed with opioid antagonists like the intravenous administration of Naloxone.

6.2.9 Lewisite—The Dew of Death

A. Introduction

Lewisite (molecular formula: $C_2H_2AsCl_3$) (blister agent), also known as Chloroethenyldichloroarsine or by code names like G-34, is a chemical compound from a chemical family called arsines. While in its pure form, lewisite is an oily, odorless, colorless liquid that contains arsenic, whereas in its impure form, lewisite is amber-to-black in appearance and has a strong odor resembling that of scented geraniums. It hydrolyses (decomposes by reacting with water) in water to form hydrochloric acid. It can persist for a long time in the liquid form in a variety of conditions, right from very hot to freezing environments.

Lewisite is a blistering agent or vesicant as it causes blistering of the skin and mucous membranes on contact. It is a complex mixture of several compounds. It is a powerful lung irritant. Lewisite breaks down quickly in water and soil. It is not found naturally in the environment.

Lewisite exposure can occur through liquid form, or by using or consuming contaminated water or food, or by breathing in the vapor form, or through skin contact or eye contact.

B. As a Chemical Weapon and Chemical Terrorism Threat

Lewisite is easy to manufacture, stable in a variety of conditions and is fatal in as little as ten minutes when inhaled in high concentrations. It can be released as a liquid into food or water supplies, or it can be vaporized and dispersed into the air. It can also be mixed with other chemical agents to increase the toxic effects. Its usefulness as a weapon of mass destruction has always been a topic of debate and controversy.

Lewisite was initially discovered by accident by a graduate student and priest while pursuing doctoral work at Catholic University of America in

1903. He was studying the reaction of the gas acetylene and arsenic trichloride in the presence of aluminum chloride. He had to abandon further investigations as he was hospitalized for severe toxic effects. Father Julius Arthur Nieuwland described this reaction in his 1904 dissertation. This toxic substance later became known as lewisite. Lewisite for the battlefield was synthesized near the end of the World War I. It is named after the American soldier and chemist, Winford Lee Lewis, who headed the team that produced lewisite based on the thesis of Father Julius Arthur Nieuwland in an effort to counter the German success with mustard gas. Lewisite was developed into a secret weapon at Nela Park, Cleveland, Ohio, and given the name "the new G-34" to confuse its development with mustard gas. In reality, German chemists had considered lewisite in early 1918 and rejected it because of hydrolysis. It was discovered too late to be used during the World War I and moreover, the vessel carrying the lewisite shipment sank while crossing the Atlantic at the cessation of the war, but research on lewisite continued in the 1920s as the "Dew of Death."

After the World War I, lewisite was sensationalized by the press thereby touting America's capability of producing deadly weapons. During the interwar period, lewisite's value as a chemical warfare agent continued to be debated. Lewisite had the *military symbol of M1* up into the World War II, thereafter, it was changed to L. During the World War II, the United States produced approximately 20,000 tons of lewisite, testing it on soldiers and secretly dropping it from planes, and keeping it on hand primarily as an antifreeze for mustard gas or to penetrate protective clothing in special situations. The field trials did not prove to be effective due its characteristic odor and lachrymatory effects that forced troops to put on masks and avoid contaminated areas, and the hydrolysis factor (lewisite breaks down when it comes into contact with water or water vapor) did not make lewisite an effective weapon in humid conditions. Lewisite was declared to be an obsolete weapon in the 1950s.

In the post-war period, lewisite production took place in the Soviet Union, United States, United Kingdom, Japan, Italy and France. During the World War II, the Soviet Union, United Kingdom, United States, Japan and Germany had a considerable stockpile of lewisite, but after the war, it was considered obsolete due to the discovery of a powerful and inexpensive antidote called Dimercaprol, also called British Anti-Lewisite (BAL). Most stockpiles were neutralized with bleach and dumped into the Gulf of Mexico. As of January 2008, the last of the global stockpile was destroyed. Despite these efforts, in 2010, lewisite was found in a World War I weapons dump in Washington, DC. In 2008, workers installing groundwater monitoring wells at the Rocky Mountain Arsenal suspended work after air monitoring results showed low levels of lewisite. This work was being done in a restricted area of the installation where chemical agents were disposed of several decades earlier. Such events definitely point a finger at the efficiency of billion-dollar disposal and cleanup operations.

Although it is generally believed that lewisite has not been used on the battlefield, Japan may have used lewisite as a weapon against China in the 1930s during the Sino-Japanese conflict. It may have also been used recently by Iraq against Iran, but it was never proved to be present in the munitions examined.

An agreement was made in the late 1990s between China and Japan to destroy the lewisite stockpiles by 2007. In mid-2006, China and Japan were negotiating disposal of lewisite in northeastern China left by the Japanese forces during the World War II. Several Chinese people have died due to accidental exposure to this toxin.

Although lewisite, a WWI weapon, after so many years, the presence of the same (lewisite residue, munitions) in civilian areas like Spring Valley, USA (during the World War I, adjacent to this area, was the headquarters for research and testing that was performed by the Chemical Warfare Service), indicates that the risk to human and animal life in lewisite-contaminated areas is difficult to gauge even today The lethal legacy of lewisite lives on.

A WORD OF CAUTION

Similar toxic conditions are believed to exist in other areas like Willoughby (USA), Volgograd (Russia), etc. Such lewisite-contaminated areas are a source of possible terror threats.

C. *Symptoms and Management*

There is no standard test to determine whether one has been exposed to lewisite. The presence of arsenic in urine sample can be used as an indicator of exposure.

The health effects and symptoms depend on the route, amount and length of exposure. The effects are similar to those of arsenic poisoning. Immediate symptoms are eye irritation, nausea, weakness, shortness of breath and headache. Severe and extensive eye exposure can result in a searing sensation, serious damage to eye parts, inflammation of the cornea, eyelids and conjunctiva, and blindness. Severe and extensive inhalation of lewisite vapor can cause severe pulmonary edema and acute damage to the respiratory tract. Very high concentrations can cause death within 10 minutes. Severe and extensive skin exposure can cause blisters and chemical burns. In the case of large, deep, arsenical vesicant burns, there may be considerable necrosis of tissue and gangrene. If lewisite enters the bloodstream, it can cause damage to the blood vessels and bone marrow.

Speed is the key—immediate decontamination of the eyes with water will prevent adverse effects. Immediate medical attention and protective eye gear is also necessary. In the case of skin contact, the exposed areas and clothing should be decontaminated with a bleach solution. Chemical burns need to be treated with standard burn therapy. In the case of ingestion, vomiting should not be induced, or fluids should not be given, or an attempt to neutralize should not be done, instead the victims should be given immediate medical attention. Medical personnel should don protective gear while handling victims. Moving away from the area of release and the agent source is the best way of protecting oneself. If lewisite exposure occurs outdoors, then the best way to avoid it is by moving to highest ground possible as lewisite vapor is heavier than air it will settle down in low-lying areas. Intravenous fluid administration is needed for low blood pressure or shock. In the case of apnea, artificial respiration should be administered. Immediate medical care, life support, symptomatic and supportive measures, and antidotes like Dimercaprol and Unithiol help to minimize the effects of lewisite exposure.

6.2.10 Methyldichloroarsine—The First Organoarsenic Compound to be Weaponized

A. Introduction

Methyldichloroarsine (molecular formula: CH_3AsCl_2) (blister agent), also known as MD or Dichloromethylarsine, is organoarsenic compound. It is a colorless volatile liquid. It reacts dangerously with oxygen and other oxidizing agents. Flames or heat can ignite MD. It has an explosive reaction with chlorine. As MD is more volatile than water, its persistence is relatively short; it will dissipate after a short time.

Modes of exposure—inhalation of vapors, absorption through the skin (dermal exposure), eye contact and ingestion. MD is highly toxic by inhalation, skin exposure or ingestion. It can be fatal if inhaled, absorbed through the skin or ingested in higher concentrations.

B. As a Chemical Weapon and Chemical Terrorism Threat

Even if MD is not that persistent (persistence of up to 3 days under normal conditions and up to 12 days in dry and warm environments), it is still lethal in nature; hence suitable as a military weapon. It can be dispersed through humidifiers, aerosols, atomizers (an *atomizer* is a device that converts a liquid into a fine spray) and explosions. MD is capable of causing causalities within minutes of dispersal.

SPOTLIGHT

Methyldichloroarsine was the first organoarsenic compound to be weaponized.

The Germans weaponized the first organic arsenical compound, MD, during the World War I between 1917 and 1918. It was developed by the United States Chemical Warfare Service in 1918, but not used.

MD is a weapon used in military operations against enemies to impair their movement and disorient them. MD can be made more lethal by mixing it with Ethyldichloroarsine (Ethyldichloroarsine is an organoarsenic compound) thereby causing irritation and biting effects on the enemy troops.

As MD can penetrate through rubber and render some gas masks ineffective, the effects of MD are almost instantaneous when it makes contact with the skin or is inhaled. Skin exposure to as little as 500 micromilliliters of MD can lead to organ damage and poisoning. MD causes blindness and it kills within minutes when the concentration level is high.

C. Symptoms and Management

MD is irritating to the respiratory tract and is likely to cause lung injury. When inhaled, MD is extremely dangerous; it can cause systemic poisoning and lung illnesses. In the case of skin exposure, symptoms like painful burning, itching and irritation with vesicle formation can occur. Prolonged skin exposure to MD liquid can lead to bone marrow destruction. The absorption of sufficient vapor or liquid through the skin can lead to blistering, systemic poisoning or death. Eye exposure can cause involuntary muscle contractions of the eyelids (blepharospasm), severe burns and permanent damage.

Protective clothing and equipment is needed to avoid bodily contact. Rescue workers and health care professionals should wear splash-resistant goggles, activated charcoal filters, suits, gloves and face shields while handling victims. All exposed material should be kept away from sources of ignition. Victims should be moved to an area with fresh air. Decontamination can be done with caustic soda and bleach. Decontamination runoff should be prevented from leaving the area of exposure or else it may cause pollution and contamination of water resources.

6.2.11 Nitrogen Mustards—A Combination of Good and Bad

A. Introduction

Nitrogen Mustards (blister agent), also known as Mechlorethamine Hydrochloride or Mustine, are structurally similar to sulfur mustards.

However, exposure to nitrogen mustards may be more immediately toxic than exposure to sulfur mustards. Nitrogen mustards are not found naturally in the environment. There are 3 variants: HN1, HN2 and HN3.

They are cytotoxic (anti-cancer) chemotherapy agents (used in cancer treatment for killing cancer cells, for example, palliative chemotherapy in lung and breast cancers). They are alkylating agents that affect DNA in the cancer cells; the cancer cells are unable to replicate and they die. Nitrogen mustards are used in combination with other drugs to treat cancers. There are several nitrogen mustards drugs like bendamustine, chlorambucil, etc., used in chemotherapeutic treatment.

There are various forms of nitrogen mustards. They can have a fruity, soapy, fishy or musty odor. They can be a solid, vapor or an oily-textured liquid. HN1 is a colorless to pale yellow oily liquid and has a faint fishy or musty odor. HN2 is a colorless liquid with a soapy odor at low concentrations and a fruity odor at higher concentrations.

Modes of exposure—contaminated water is consumed or comes in contact with the skin, exposure to the vapor form through inhalation, eye contact or skin contact, and exposure through direct contact with the liquid form.

B. As a Chemical Weapon and Chemical Terrorism Threat

Nitrogen mustards were produced in the 1920s and 1930s as potential chemical weapons. These agents were synthesized in the late 1930s. The weapon designation for them is HN1, HN2 and HN3. HN1 was originally developed to treat warts but later was identified as a potential weapon. HN2 was designed as a military agent but was later used in chemotherapy; HN2 became a prototypical cancer chemotherapeutic compound. HN3 was designed solely as a military agent and is the only one of the nitrogen mustards that is still used for military purposes. There are no verified reports of their actual use as chemical weapons.

The Germans and Americans started the military production of nitrogen mustards agents in 1941 and 1943, respectively. Germany considered nitrogen mustards as a great area denial weapon. During the World War II, nitrogen mustards were studied at Yale University as a possible drug for cancer treatment. Concurrently, during the same period (during the World War II), a German *air* attack on the harbor of *Bari, Italy,* struck a US ship loaded with mustard agent munitions and due to this raid several hundred people were accidentally exposed to mustard gas. Studies done on the survivors showed a decreased number of lymphocytes; the survivors were found to have very low white blood cell count. The history of cancer chemotherapy is traced to the Bari incident. These studies made researchers look for other substances that might have similar effects against cancer. The Bari incident and the work of the Yale University researchers led to the post-war development of the first chemotherapy drug Mustine (HN2). Because of wartime secrecy restrictions, it was not until 1946 that all the developments and results were published openly. On the one hand, sulfur mustard was used extensively in the World War I and World War II, and on the other hand, its relative nitrogen mustard was stockpiled by several countries during the World War II, but it was never used on the battlefield.

HN3 is the most likely weapon of choice on the battlefield because of its persistence in the field, ability to harm the mucous membranes, skin, eyes and lungs, and stability. Nitrogen mustards have the potential to be used as chemical terrorism agents because of their extreme vesicant effects. Terrorists are more likely to use the HN1 variant as its lethal dose through inhalation is substantially smaller in comparison to sulfur mustard.

C. *Symptoms and Management*

The extent of poisoning and symptoms depend on the route, amount and length of exposure. Unfortunately, nitrogen mustards do not only kill cancer cells, but any other cell that grows and divides thereby reducing the white blood cell count, platelet count, red blood cell count; this cell damage can cause anemia or infection. Exposure affects the eyes, skin, immune system, bone marrow and respiratory tract. Low exposure can cause pain, irritation and swelling of the eyes, nausea, abdominal pain and mouth sores, whereas high exposure can cause burns and blindness. The skin can experience redness and blistering. Symptoms and signs related to the respiratory tract include sore throat, shortness of breath, cough and nose pain. Bone marrow suppression can also occur. High exposure can cause seizures, chronic respiratory diseases and tremors. Repeated or prolonged exposure may cause leukemia.

One should quickly move away from the area of the release to an area where fresh air is available. As nitrogen mustard vapor is heavier than air it will hover near the ground, therefore, the best way to avoid exposure to the chemical release is by moving to the highest possible ground. Treatment includes supportive medical care and removal of nitrogen mustards from the body, clothing and other items. Clothing and other items should be sealed within a plastic bag which should be placed in another plastic bag. Contaminated clothing, objects and victims should be handled with the help of a protective ensemble. Affected areas of the body should be washed with water and soap. Eyes should be thoroughly cleansed with water. There is no standard antidote available.

6.2.12 *Phenyldichloroarsine—A Variant of Lewisite*

A. Introduction

Phenyldichloroarsine (molecular formula: $C_6H_5AsCl_2$) (blister agent) also known as Dichlorophenylarsine or PD, is a colorless, odorless organic arsenical vesicant and vomiting agent with an oily texture. It is a vesicating *analog of lewisite*. Its blistering effect is similar to lewisite, but less severe. In the impure form, PD may have a slight brown color and an unpleasant odor. When it reacts with moist air or water, PD forms toxic hydrochloric acid. PD is incompatible with bases and acids. It is a dangerous fire hazard and a flammable liquid.

SPOTLIGHT

Phenyldichloroarsine has been used both for commercial (solvent, organic synthesis) and military purposes (a military poison gas, vesicant and blister agent).

One can get exposed to this chemical through ingestion, inhalation, or absorption through the skin or contact with the eyes. PD is toxic by inhalation and skin absorption.

B. As a Chemical Weapon and Chemical Terrorism Threat

PD can be mixed with mustard agents for use in chemical warfare. PD is useful as a vomiting agent in open spaces, but it is highly effective in closed environments like caves, trenches and basements. PD was developed for use in wet environments because of its tendency to persist in cool and shaded areas. Although PD's vesicant effects are mild and delayed, it can be used to incapacitate troops in field operations.

PD was first used as a war gas in a limited manner in the World War I and it proved to be less effective than other vomiting agents. It was developed by France and Germany for use as a chemical warfare agent during the World War I and by Germany during the World War II.

As warfare weapons, MD and PD were operationally less effective than lewisite. Moreover, PD is easily recognizable in the field due to which it is not as effective as other blister agents. PD is now considered as an obsolete chemical warfare agent.

PD has been used in police operations. PD grenades have been used by high-security setups like banks to deal with security breaches.

C. Symptoms and Management

PD is a strong irritant to the eyes, skin and tissue. PD exposure has an immediate effect on the eyes and a delayed effect (30 minutes to 60 minutes) on the skin. High doses can result in blindness. Long-term exposure can cause systemic poisoning and bone marrow destruction. Ingestion can cause severe irritation or burns of the stomach and mouth. Inhalation can cause lung irritation, cough and shortness of breath. Severe inhalational exposure can cause pulmonary edema. PD exposure can also damage the nervous system and liver.

The immediate response is to move victims away from the site of exposure to fresh air. If eye or skin exposure has occurred, then contaminated clothing and other objects should be removed immediately and the affected areas must be flushed with lukewarm water for at least 15 minutes. Acute exposure may require life support for the victims. Full protective ensemble—clothing, face shield, goggles, gloves and masks are useful in minimizing the splash hazard. Emergency procedures of artificial respiration are needed in case of breathing difficulties. Local exhaust ventilation at the site of exposure needs to be used. Decontamination runoff should be prevented from leaving the area of chemical release or else it may cause pollution and contamination of water resources. PD should be prevented from entering basements, waterways or confined areas.

6.2.13 Sarin—The Poor Man's Atomic Bomb

A. Introduction

Sarin (molecular formula: $C_4H_{10}FO_2P$) (nerve agent), also known as GB or Phosphonofluoridic Acid, is a manmade chemical warfare agent. It is not found naturally in the environment. It is a colorless, non-persistent liquid, which gives off no odor when vaporizing. The vapor form is colorless. It is a lethal cholinesterase inhibitor (cholinesterase is an enzyme/substance that helps the nervous system work properly). Sarin disrupts the functioning of the nervous system by inhibiting the cholinesterase enzyme. *Cyclosarin* (molecular formula: $C_7H_{14}FO_2P$), also known as GF or Cyclohexyl Methylphoshonofluoridate, is a highly deadly chemical warfare agent. Like its predecessor *sarin*, cyclosarin is an organophosphate nerve agent, but it has different characteristics from sarin. Unlike sarin, cyclosarin is flammable and is a persistent liquid (evaporates relatively slowly, about 69 times slower than sarin).

Modes of sarin exposure—contaminated water is consumed or comes in contact with the skin, exposure to the vapor form through inhalation, eye contact or skin contact, touching contaminated surfaces, and exposure through direct contact with the liquid form.

B. As a Chemical Weapon and Chemical Terrorism Threat

Sarin was originally developed in 1938 in Germany as a pesticide. Other nerve agents like Tabun, Cyclosarin and Soman were also developed through the German research efforts during the World War II. Germany had produced a considerable amount of sarin munitions, but these munitions were not used in battle, only hauled and shipped from one warehouse to another. Research on sarin was carried out at a pilot plant in *Dyrenfurth*, Germany. Although sarin was produced at this plant, it was never used during the World War II. The Soviet Army captured this plant at the end of the war and restarted sarin production in 1946. The United States produced sarin between 1950 and 1956.

Sarin is called the poor man's atomic bomb because of the large number of people that can be killed by a small amount. According to experts, sarin is 500 times more toxic than cyanide. A fraction of an ounce of sarin on the skin can be fatal.

A WORD OF CAUTION

A very small dose of sarin (a pinprick-sized droplet) is needed to kill humans. It kills by effectively crippling the nervous system.

The gas or liquid form is highly volatile and lethal. As sarin is the most volatile of the nerve agents, it can quickly evaporate from a liquid into a vapor thereby exposing people to the vapor form; people experience intoxication without coming in contact with the liquid form. As sarin evaporates at room temperature, this characteristic makes it ideal for aerosolized dispersal and people will not know of their exposure until they experience symptoms as sarin is odorless. A person's clothing can release sarin for about 30 minutes after it has come in contact with sarin vapor, and this in turn can lead to exposure of other people thereby causing more damage and causalities, both in the military and non-military settings. Humid and wet conditions deteriorate sarin, but as the temperature rises up to a certain point, its lethal duration increases.

Iraq used sarin against the Kurds during the 1987-88 Anfal campaign (the Kurdish Genocidal Campaign). Iraq used sarin and cyclosarin mixture during the Iran-Iraq war (1980–1988) as this mixture proved to be a more persistent chemical agent in the dry hot desert climate of Iraq. Controversial recent discoveries (2004) of sarin and mustard gas munitions in Iraq have created a whirlpool of questions related to the manufacture, possession and use of banned chemical weapons. Now whether there is much more than what meet the eyes is a matter of intelligence speculation. Some Iraqi scientists believe many similar weapons could be discovered that were stockpiled by the former regime (they were either buried underground or transported to Syria) and also could be brought back by insurgents. They could also be sold on the black market and there is a possibility that they could be used outside of Iraq.

Sarin was used in two terrorist attacks in Japan in 1994 and 1995—the

Japanese cult Aum Shinrikyo killed a dozen people and injured thousands in multiple sarin attacks in the Tokyo subway system in 1995. This deadly act sent shock waves throughout the world. This cult had also injured and killed people in a sarin gas attack in the central Japanese city of Matsumoto in 1994. Though the number of causalities was low as a low-lethality batch of sarin was used yet these two attacks managed to cause extreme panic in the public. There is no doubt that terrorists can use sarin to create this kind of panic and terror among civilians. For terrorist groups and rogue states, sarin can be manufactured in secret laboratories as it does not require special equipment. On the downside, sarin is very dangerous to handle because sarin as a weapon can be difficult to mix properly and safely, and can also be highly unstable.

C. Symptoms and Management

The degree of symptoms depends on the amount, length and route of exposure. The onset of symptoms occurs within a few seconds after exposure to the vapor form and within a few minutes to 18 hours after exposure to the liquid form.

Common symptoms include nausea, muscle twitches, tightness in the chest, tremors, diarrhea, headache, nasal congestion, constriction of pupils and abdominal cramps. Low-level to moderate-level exposure can cause severe eye and skin irritation, drowsiness, excessive sweating, abnormal blood pressure level, and slow or fast heart rate. Exposure to a large dose can cause severe symptoms like respiratory failure, incapacitation, convulsions and even death.

As sarin vapor is heavier than air, it will sink to low-lying areas; hence it is best to move to the highest possible ground to avoid exposure hazard. Protective gear like protective masks should be used. Decontamination of areas and surfaces exposed is a top priority. Liquid sarin needs to be washed off from exposed body parts with large amounts of soap and water. For respiratory problems, oxygen delivery systems should be used. In the case of constricting pupils, immediate medical attention and decontamination of the eyes is needed. Immediate antidote therapy and ventilator support for severe exposure is crucial. Antidotes like atropine are used. Antidotes are most effective if given as soon as possible after the exposure.

6.2.14 Sulfur Mustards—A Major Military and Civilian Threat

A. Introduction

Sulfur Mustards (molecular formula: $C_4H_8Cl_2S$) (blister agent), also known as Mustard Gas, Yellow Cross Liquid or Yperite, are a class of related cytotoxic and vesicant chemical warfare agents. They are not found naturally in the environment. They can exist as a solid, gas or liquid. Their odor resembles like that of horseradish, garlic or mustard. Until recently, they were available for treating psoriasis (a skin condition). Currently they have no medical application. They are a blister/vesicant agent as they cause blistering of the respiratory tract, mucous membranes and skin on contact. They damage the DNA. They are a carcinogen—a cancer-causing agent. Under normal weather conditions, sulfur mustards can persist for 1-2 days, and in cold conditions, they can last from weeks to months.

Modes of exposure—by consuming contaminated water or getting sulfur mustards on the skin, exposure by coming in contact with the liquid form, and exposure to the vapor form through inhalation, skin contact or eye contact.

B. As a Chemical Weapon and Chemical Terrorism Threat

Factors like ease of production, lack of effective protection and treatment, problems with decontamination, and exposure is not always detected immediately, especially in skin exposure due to the latent and sign-free period make sulfur mustards a significant threat.

During the World War I, sulfur mustards was introduced as a chemical weapon (as Levenstein mustard) by the Germans. Approximately 80% of chemical weapons-related casualties in the World War I were caused by sulfur mustards. Once this agent proved its success, Germany's production of sulfur mustards went into overtime. At that time, this agent was preferred to other chemical warfare agents, because of the fact that this agent could penetrate all protective materials and masks that were available at the time. The British used their first sulfur mustard aerosol weapon in 1918. During the World War II, this agent was used by the Japanese Imperial Army against the Chinese.

Only 10 mg in the lungs is required to kill a person. Sulfur mustards when used in the pure form are colorless and odorless, and therefore people will not know of exposure until they begin to experience symptoms. Sulfur mustards in the liquid form can be dispersed by a crop spraying aircraft.

As sulfur mustards can persist somewhere from 1 to 2 days to weeks and months depending on weather conditions, makes sulfur mustards a good candidate for chemical warfare. But it would take a large amount of sulfur mustards to kill many people. It would be best used for causing more lasting injuries.

A WORD OF CAUTION

It is possible to increase the persistency of sulfur mustards by dissolving them in non-volatile solvents. As a result of which, a thickened agent is obtained, which is very difficult to remove.

Presently, small-scale production of sulfur mustards is done for defense research purposes. Bulk production is not undertaken. Large amounts of sulfur mustards have been disposed of at sea. Sulfur mustards may also be found buried at several defense facilities. Sulfur mustards are still considered as a potential military threat and terrorist threat.

C. Symptoms and Management

The severity of health effects and symptoms depends on the length of exposure, amount of dose and route of exposure, environmental temperature, and the sensitivity of people to sulfur mustards.

High exposure to the vapor form can cause chronic respiratory infection and even death. When inhaled, sulfur mustards can cause sinus pain, injury to the mucous membranes of the respiratory tract, cough, sneezing and shortness of breath. In the case of heavy inhalational exposure, death can occur due to pulmonary edema or mechanical asphyxia. When ingested, sulfur mustards can cause nausea, abdominal pain, diarrhea, vomiting and fever. The eyes can experience pain, swelling and irritation; harsh exposure can cause blurred vision or even blindness. The skin can experience itching, redness and blistering, and in severe exposure, necrosis of the epidermal cells extends to the underlying tissues as well. Long-term exposure can cause serious symptoms like respiratory and lung infection, second or third degree burns, scarring, cancer and even death.

In case of exposure, quick decontamination is needed. Medical professionals and rescue personnel should be equipped with a full protective ensemble to avoid secondary exposure. As sulfur mustards are heavier than air, they tend to accumulate in low-lying areas, due to which the best way to avoid exposure is to move to higher ground. Rapid decontamination of this chemical is necessary to avoid tissue damage. Chemical inactivation using chlorination is effective against sulfur mustards. In low-dose skin exposure, a paste containing a chlorinating agent like chloramine can be used to protect the skin. No standard treatment or antidote exists for sulfur mustards; hence supportive medical care and prophylactic measures are needed to minimize the symptoms of exposure.

6.2.15 Tabun—The First of the G-Series Nerve Agents

A. Introduction

Tabun *(molecular formula:* $C_5H_{11}N_2O_2P$*)* (nerve agent), also known as GA or Ethyl dimethylphosphoramidocyanidate, is a manmade chemical warfare agent classified as a nerve agent. It is not found naturally in the environment. Tabun is the first of the G-series nerve agents.

When heated, tabun turns into the vapor form. In the pure form, tabun is a colorless, tasteless liquid with a slightly fruity odor, whereas the industrial product has a brownish color and has an odor similar to bitter almonds. Tabun is more volatile than VX but less volatile than sarin.

Exposure routes—by consuming contaminated water or food, or coming in contact with contaminated water or food, through skin contact or eye contact, or through inhalation of tabun mist or vapor.

B. As a Chemical Weapon and Chemical Terrorism Threat

Tabun was the first nerve agent discovered. Tabun was accidentally discovered in 1936 by the German researcher, Dr. Gerhard Schrader of IG Farben, Germany (IG Farben was a German chemical industry conglomerate, which was involved in numerous war crimes during the World War II), during his research with a class of compounds called organophosphates for creating a more effective insecticide. Instead of a new insecticide, Dr. Schrader discovered tabun, a chemical extremely toxic to humans as well as insects. Industrial production began in 1942 at a special factory in the town of *Dyrenfurth*. At least 12,000 metric tons were manufactured between 1942 and 1945.

Tabun is considered among the easiest of the nerve gases to manufacture. Tabun (along with other never agents) was used as a nerve gas in chemical warfare. Iraq used tabun during the Iran-Iraq war; this is the only known battlefield use of nerve agents. In 1984, Iraq became the first nation to use a nerve agent on the battlefield when it used Tabun-filled aerial bombs during the Iran-Iraq war. Intelligence analysts believe that many countries have the technology to manufacture nerve agent munitions, including tabun. As tabun is easy to produce, and with a mini laboratory equipped with protective gear and apparatus, it is possible for terrorists to manufacture small quantities of tabun for use in terrorist acts.

Tabun is an immediate but short-lived threat because of its high volatility. Although tabun has a faint fruity odor, the odor may not be detectable, therefore, victims may not know of the exposure hazard until they begin to experience symptoms. As tabun mixes easily with water, it can be used to poison water sources. Even a tiny drop of tabun on the skin can cause muscle twitching and sweating. Contaminated clothing can release tabun for about 30 minutes after contact with the vapor form, which can lead to exposure of other people. If victims do not receive immediate treatment in the case of a lethal dose, death will occur after about 20 minutes. Victims who do not get exposed to a lethal dose but also do not receive immediate medical treatment may suffer permanent neurological damage.

C. Symptoms and Management

The health effects and symptoms depend on the length of exposure and degree of intoxication. The onset of symptoms occurs within a few seconds after exposure to the vapor form and within a few minutes to 18 hours after exposure to the liquid form. Symptoms appear much more slowly from skin exposure than from respiratory exposure.

In the case of low to moderate level concentrations, tabun can cause blurred vision, watery eyes, runny nose, rapid breathing, tightness in the chest, cough, nausea, vomiting, diarrhea, headache, weakness, and disturbance in the heart rate and blood pressure. Severe cases may lead to paralysis, convulsions, respiratory failure and even death. As tabun breaks down slowly in the body, repeated exposures can have a cumulative effect.

For ingestion exposure, the Nerve Agent Antidote Kit, Mark I (This kit consists of two autoinjectors containing Atropine Sulfate and Pralidoxime Chloride) should be administered. Skin exposure needs to be treated with copious irrigation of the affected skin areas (lots of *water and soap, sodium carbonate* or bleach*)*. Contaminated clothing and other objects should be sealed in a plastic bag. Then the first plastic bag should be sealed in a second plastic bag. Supportive care, rapid decontamination, moving to an area with fresh air, antidotes, protective gear for handling victims and oxygen administration for breathing difficulty facilitate the process of recovery from exposure.

6.2.16 VE and VR— Closely Related to the Better-Known VX Nerve Agent

VE (molecular formula: $C_{10}H_{24}NO_2PS$) (nerve agent) is a "V-series" nerve agent. It is closely related to the better-known VX nerve agent. VE has similar lethal dose levels, symptoms and treatment as that of VX. However, this second-generation nerve agent is known to cause coma in addition to standard seizures. Moreover, as the window for effectively treating second generation V-series seizures is shorter, administration of standard nerve gas antidote pralidoxime becomes ineffective unless it is given immediately after the exposure. The second-generation" V series agents are believed to be a product of a cold war era Russian chemical weapons development program

VR (molecular formula: $C_{11}H_{26}NO_2PS$) (nerve agent), also known as Soviet V-gas, Russian VX or R-33, is closely related to the better-known VX nerve agent. VR is a structural isomer of VX. VR was investigated and developed by the Soviet Union's Scientific Research Institute No. 42 during the 1950s. This development paralleled the development of VX in the west. Soviet research was partly based on the knowledge of the molecular formula of VX. Although they knew of the VX molecular formula, the Soviet Union continued with extensive research work on VR. It is believed that the initial study of VR had already been undertaken by the Soviet Union much before the United States settled on VX, and moreover factors like VR is difficult to treat and detect made the Soviet Union to continue research on VR. This theory is supported by the fact that a pilot production facility for VR was set up in Stalingrad in 1956, two years before VX was selected by the United States. Despite the early study and experimentation, large-scale VR production began at a facility in Novocheboksarsk after many years in 1972. Now most of the VR stockpile has been destroyed.

6.2.17 VX—The Most Potent of All Nerve Agents

A. Introduction

VX (molecular formula: $C_{11}H_{26}NO_2PS$) (nerve agent), also known as Methylphosphonothioic acid or by the UK Rainbow Code *Purple Possum*, is a manmade chemical warfare agent classified as a nerve agent. VX is not found naturally in the environment. It is the most potent of all nerve agents. It is almost like motor oil in terms of texture, feel and consistency, and it is colorless and tasteless. VX is primarily a liquid at room temperature, but when heated to very high temperatures it turns into a vapor/mist. VX is an organophosphate nerve agent. It inhibits the function of the acetylcholinesterase enzyme.

VX exposure can occur through inhalation of the mist, through skin and eye contact, or through ingestion or exposure to contaminated food or water. VX is the least volatile of all nerve agents, which makes it hazardous through the dermal route and more toxic by inhalation.

B. As a Chemical Weapon and Chemical Terrorism Threat

VX is classified as a weapon of mass destruction by the United Nations. It is ten times more toxic than sarin. Moreover, as VX is tasteless and odorless, people will not know of exposure until they experience symptoms. As little as 15 milligrams of the VX gas on the skin kills the victims. Death usually occurs within 15 minutes after exposure to a fatal dosage. Contaminated clothing can be an effective weapon as it can release VX for about 30 minutes after contact with the VX vapor.

A WORD OF CAUTION

VX is an area denial weapon due to its physical properties.

VX is very persistent in the environment as it the least volatile of all the nerve agents and it is very slow to evaporate thereby it poses a long-term threat. Under average weather conditions, VX can last for days on objects that it has come in contact with. Under very cold conditions, VX can last for months. Certain forms of VX are so adhesive that it is almost impossible to remove them; hence if these forms are used in the battlefield, they are certain to make the area uninhabitable as they contaminate the buildings, ground, equipment, surfaces, etc., and thereby reduce the fighting efficiency of enemy troops.

VX was originally developed in the United Kingdom in the early 1950s. In the 1950s, the United Kingdom was searching for new insecticides and they came across compounds that were highly toxic to humans. Later on in 1955, these compounds were designated the V-series agents. VX was developed in the Porton Down Chemical Weapons Research Centre, Wiltshire, United Kingdom in 1952. Britain traded the VX technology with the United States in exchange for know-how on thermonuclear weapons. The United States then produced large amounts of VX in 1961 at the Newport Chemical Depot.

There is no conclusive evidence that V-series agents, including VX, were used on the battlefield during the World War II, but they were tested in concentration camps. It is believed that VX was used in chemical warfare during the Iran-Iraq War in the 1980s. Although, synthesizing VX is complicated and very dangerous, and needs a sophisticated laboratory setup, still the Japanese cult, Aum Shinrikyo, managed to recruit trained chemists from Japanese universities and synthesized a small quantity of VX (100 to 200 grams) to use for assassinations. The cult used it to attack three people. Two people were injured and one died, who is believed to be the only victim of VX ever documented in the world. Even if the process of producing VX is dangerous and complicated, terrorist groups who don't possess the sophistication and knowledge are quite likely to steal VX from poorly guarded chemical weapons facilities. A successful VX attack (or any other nerve agent) by terrorists will have adverse physical, economic and psychological effects.

The Dugway sheep incident, also known as the Skull Valley sheep kill, was a 1968 sheep kill in which six thousand sheep were killed near the base at Dugway Proving Ground, USA. Several reports link this incident to open-air chemical weapons testing. One of the possible causes was that during an open-air test, a faulty spray nozzle caused an aircraft to continue spraying VX as it climbed to higher altitudes. The Dugway sheep incident led to President Richard Nixon's decision to ban all open-air chemical weapons testing in 1969. Thereafter, in the late 1960s, the United States began destroying its stockpiles of chemical agents.

The only countries known to possess VX are the United States, Russia and Syria. VX production and stockpiling was outlawed by the Chemical Weapons Convention of 1993, which has led to widespread destruction of VX stockpiles worldwide. Several methods like incineration and chemical neutralization have been used to eliminate VX stockpiles. For example, the Johnston Atoll Chemical Agent Disposal System, America's first chemical munitions disposal facility used incineration, and this was followed by other facilities and installations like the Pine Bluff Arsenal, Anniston Army Depot, etc. The Newport Chemical Depot, a bulk chemical storage and destruction facility, used chemical neutralization to destroy the VX stockpile. The United States is helping Russia in destroying chemical weapons, including VX, through the Nunn-Lugar Global Cooperation Initiative.

C. Symptoms and Management

Lowered acetylcholinesterase levels are indicators of nerve agent exposure in victims. Symptoms depend on the length, amount and mode of exposure. The onset of symptoms for exposure to the liquid form occurs within a few minutes to up to 18 hours. The onset of symptoms for exposure to the vapor form appears within a few seconds.

Low-level exposure can cause symptoms like diarrhea, runny nose, blurred vision, excessive salivation, tightness in the chest, rapid breathing, drowsiness, cough, headache, weakness, slow or fast heart rate, nausea and abnormally low or high blood pressure. A high dose can cause paralysis, convulsions, respiratory failure and even death. In mild or moderate exposure, victims usually recover completely, whereas severe exposure without immediate treatment can lead to death.

Treatment includes supportive therapy, emergency and first aid procedures, removal of VX from the body and nerve agent antidote drugs like the Mark I kit injectors to minimize the effects of exposure. If exposure occurs indoor, then one should move outdoors. In the case of outdoor exposure, as VX vapor is heavier than air it settles down in low-lying areas, therefore, it is best to move to the highest possible ground. Affected areas of the body should be washed with water and soap. Eyes should be thoroughly cleansed with water. Removal or neutralization of the agent present on the skin is very important. Decontaminants should be used to ensure the removal of the agent and any toxic degradation products. Contaminated clothing, objects and victims should be handled with the help of protective equipment. Specific symptoms like convulsions can be treated with a sedative/antiepileptic drug like Diazepam. Oxygen or oxygen-enriched air should be used for ventilation. Intravenous infusion of anti-nerve agents such as Atropine Sulfate and Pralidoxime Salts is helpful in restoring the acetylcholinesterase activity and reversing the damage.

6.2.18 White Phosphorus—From the War Trenches to Modern-Day Genocide

A. Introduction

White Phosphorus (P_4), also known as WP, is a colorless to yellow translucent waxy fat soluble substance, which is highly reactive with oxygen. It is produced from phosphate rocks. It emits smoke. It glows in the dark and this property of WP is used in tracer bullets (a tracer bullet is a bullet whose trajectory can be observed by a trail of smoke). WP is pyrophoric (spontaneously flammable) in nature, and its fire can be difficult to extinguish. It can burn combustible items upon contact.

WP is used in the production of phosphoric acid and phosphate-based fertilizers. It is used to make cleaning compounds, fertilizers and additives.

Exposure routes—inhalation, ingestion of contaminated food or water, occupational exposure or coming in contact with contaminated soil.

B. As a Chemical Weapon and Chemical Terrorism Threat

Apart from non-military uses, WP is also used for military purposes in the form of incendiary, smoke screen, flare and gas. It is known as "Willie Pete" in military jargon.

In the military settings, WP is used for signaling, screening and incendiary purposes. For example, WP can be used to limit the enemy's visibility and the ability to function by producing smoke, or to limit the enemy's vision, or to flush enemies from their tunnels and fox holes into more open areas where standard ammunition can be used on them, or to ignite ammunition, fuel supplies and other materials, or to jam weapon tracking devices, or in target location and navigation, or to protect potential targets, or to camouflage movement.

The history of WP has proven its lethal potential in all settings, right from the war trenches to modern-day genocide. WP's reaction with elements like oxygen, hydrogen and phosphorous produces various deadly effects like blinding light when a grenade explodes (as the oxygen reacts with the phosphorus causing the phosphorus to instantly burn producing blinding light). WP is considered as one of the worst chemical weapons. Many of the WP victims during the World War II were shot by German troops to end their suffering.

WP was first used by Fenian arsonists in the 19th century, best described as the "Fenian Fire." They used it in the form of a solution of WP in carbon disulfide and this mixture was referred to as the Fenian Fire.

The British Army introduced the first factory-built WP grenades in 1916. The British troops used WP bombs against Kurdish villagers and Al-Habbaniyah in Al-Anbar province during the Great Iraqi Revolution of 1920. WP was used as an incendiary weapon by British, American and German forces. It was also extensively used during the World War II in the form of marker shells, colored flares, smoke screens, incendiaries, etc. In the battle for Grozny in Chechnya, every fourth or fifth Russian artillery or mortar round fired was a smoke or white phosphorus round.

Recent WP attacks have sparked up global interest and concern over WP—Israel used white phosphorus wedges, canister liners and shells over densely populated areas of Gaza during its recent Operation Cast Lead in 2009 against Hamas targets. The indiscriminate use of WP by Israel Defense Forces was condemned worldwide and reports of the Human Rights Watch stated that Israel used WP in violation of the laws of war as civilians suffered and died for no fault of theirs and also a lot of civilian structures were damaged. During the 2006 Israel-Lebanon conflict, Israel claimed to have used WP shells against enemy troops, whereas Lebanon claimed that WP was used against civilians. In another controversial attack, the US troops used WP bombs against enemy combatants in the Iraqi city of Falluja in 2004 in shake-and-bake missions. As recently as 2011, the United States and other international forces in Afghanistan were using WP against Taliban rocket positions. There are alleged reports of WP shells used during the NATO operation in Libya in 2011.

As WP is lethal in all states (gas, liquid and smoke) and toxic by all routes of exposure, it is a potential chemical warfare and chemical terrorism agent. As WP munitions are stocked by many countries and most military units, it is possible for terrorists to steal from such stocks and use it for their operations against civilians.

SPOTLIGHT

Over the years, several countries have expressed concerns about incendiary weapons, including WP, and have also shown the willingness to reexamine existing protocols and conventions to put constraints on the same.

C. Symptoms and Management

Ingestion causes diarrhea, vomiting, nausea, and pain in the throat and abdomen. Inhalation causes headache, coughing, and irritation of the lungs, respiratory tract and throat, and in severe cases of prolonged exposure like occupation-related inhalation, a condition known as "phossy jaw" can be experienced where delayed wound healing of the mouth and necrosis of the jaw occurs. When WP comes in contact with the skin, it creates intense and persistent burns and also can lead to tissue damage by blocking off blood circulation in the affected areas. WP has rapid dermal penetration; it is like feeding off of the oxygen in the body as a fuel.

Skin exposed to WP needs prompt wound irrigation in order to decrease the extent of dermal injury. The affected areas of the skin need to be treated with saline-soaked or water-soaked pads that must be kept continuously moist so that the affected areas do not become dry as WP can re-ignite. Tetanus prophylaxis treatment needs to be administered as a part of effective burn therapy. Residual WP particles need to be removed surgically and topical antimicrobial burn creams need to be used post-surgery. Contaminated clothing and other objects should be removed to prevent re-ignition from contact with air and to avoid burning through to the skin. Decontamination runoff should be prevented from leaving the area of exposure or else it may contaminate water resources. Health care providers should avoid contact with WP by using protective clothing and equipment.

6.3 Pictographic Recap of Chemical Weapons

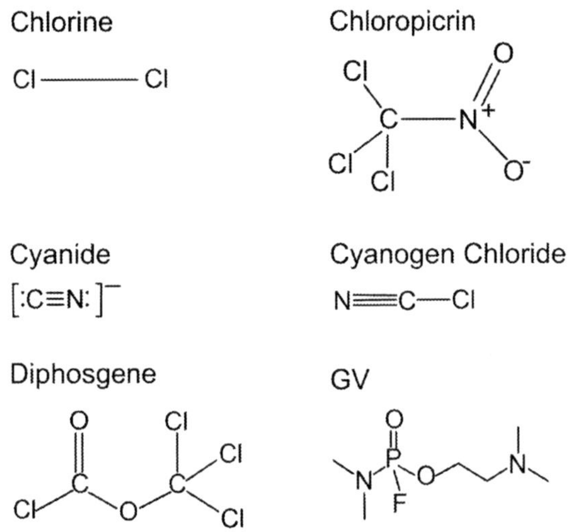

Figure 6.1: Chemical Weapons Part 1

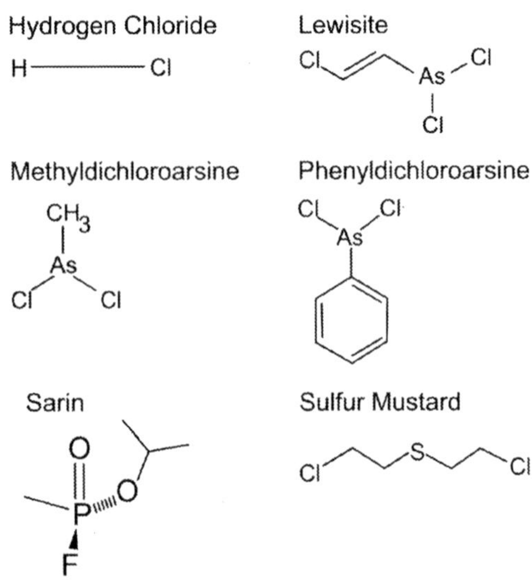

Figure 6.2: Chemical Weapons Part 2

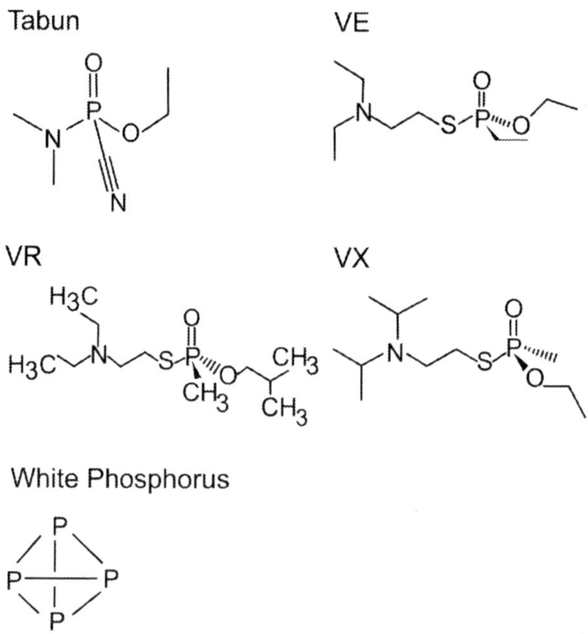

Figure 6.3: Chemical Weapons Part 3

6.4 The Final Word—The Butterfly Effect

Experts have predicted that every potential military antagonist, large nation or small nation, state actor or non-state actor, will continue to acquire, threaten with or use all forms of WMD. As new sources of WMD technologies continue to be more available around the world, the challenge of curtailing new WMD becomes difficult. Even non-state actors will become more interested in new WMD as they offer a relatively inexpensive way of inflicting serious damage on a large scale.

As these new types become more and more accessible, there is bound to be considerable overlap and dual use of these new weapons, and this means it is definitely the time to shift policy focus from "CBRN" to "CBRN + Modern-day scenarios liker Cyber Weapons and Economic Warfare," and even more types like nanotechnology, agricultural terrorism, cybernetics, weather warfare, etc. This essentially means that not only are our armed forces at risk, but even the civilian populations and the cyber world are in grave danger. Civilian and military authorities need to have a complementary plan of action in place, and they need to constantly keep each other informed of the latest activities and developments in this field.

It is true that constructive applications of new technologies will develop alongside destructive applications, but the number of destructive uses is always going to be higher. There is no magic pill against these new threats. The political role for preventing proliferation of these new weapons needs to be augmented even further with a holistic approach. The necessary precautionary measures, civil defense resources, awareness and training programs, state-of-the-art equipment, defensive countermeasures, and conscientious efforts from concerned citizens, government leaders, scientists, engineers, military leaders, export industry, and expert groups in the fields of policymaking, military strategy and technology all are needed to play a vigilant role in unison. All the involved stakeholders need to closely follow the new types of WMD and the delivery systems of such weapons in order to make informed decisions.

In the Nobel Peace Prize acceptance speech of 2001, the seventh Secretary-General of the United Nations, Kofi Annan, commented on a phenomenon known as the "Butterfly Effect"—according to scientists, the world of nature is so small and interdependent that a butterfly flapping its wings in the Amazon Rainforest can generate a violent storm on the other side of the planet. He noted that, for better or worse, the world of human activity also has its own "Butterfly Effect." In conclusion, human actions can either annihilate the world or save the world. *Save the world* is what we need to focus on—this is a step in the right direction, but it needs to be a giant, global step to deal with the threats of WMD effectively.

Table 6.1: Power Phrases to Deal with WMD

Engaged international community	Cohesive and coordinated approach
Control and ban import and export of dual-use materials	Growth of regional nuclear-weapons-free zones
Multilateral control and support	Disaster preparedness training
Marshal resources, both monetary and non-monetary, for combating proliferation	Public education and mobilization
Tightened verifications and safeguards	Substantial dismantlement
Better intelligence about the various proliferation scenarios	Appropriate mitigation and preemptive measures

So there you have it—*how, who, when, where, why, what* and *what next* of Weapons of Mass Destruction (WMD).

6.5 Questions for Reflection

A. How does the chemical weapons threat compare to the threats from other weapons of mass destruction?
B. What effects do sea-dumped chemical weapons have on the environment?
C. How much the general public should worry about chemical warfare?
D. Which chemical weapons pose the biggest threat?
E. What is the motivation behind developing offensive chemical weapons?
F. In what ways can the assistance, protection and cooperation between countries be improved to deal with chemical warfare?

Appendix A

FURTHER READING

Appendix A provides a snapshot of the various learning material and useful resources available for further reading, study and research. The resources throw light on critical issues like potential threats, control and disarmament issues, current and future challenges, etc. There is something useful in these resources for everyone, right from a casual reader to a committed researcher.

Books, Journals, White Papers and Scholarly Articles: The following list contains details of books, white papers, reports, essays and scholarly articles for further reading; these resources focus both on the historical background and current issues of WMD.

A. Bulletin of Atomic Scientists magazine focuses on the dangers to the survival and development of humanity from nuclear weapons and other weapons of mass destruction. It provides information on developments that impact public policy issues and global security.

B. Bradley, John, ed. 2000. *Learning to Glow: A Nuclear Reader.* Tucson: University of Arizona Press.

C. Bundy, McGeorge. 1988. *Danger and Survival: Choices about the Bomb in the First Fifty Years.* New York: Random House.

D. Cohen, Daniel. *The Manhattan Project*. New York: Twenty-First Century Books, Inc., 1999.

E. CBW Conventions Bulletin, formerly known as the Chemical Weapons Convention Bulletin, is the quarterly journal of the Harvard Sussex Program. It provides up-to-date news, a calendar of key scheduled forthcoming events, details of recent publications, detailed reports of the activities of the Biological Weapons Convention, etc.

F. Etel Solingen, *Nuclear Logics: Contrasting Paths in East Asia & the Middle East* (Princeton, NJ: Princeton University Press, 2007).

G. Frank, Richard B. 1999. Downfall: The End of the Imperial Japanese Empire. New York: Random House.

H. Gardner, Gary T. 1994. Nuclear Nonproliferation: A Primer. Boulder: Lynne Rienner Publishers.
I. *Norris, John and Will Fowler. Nuclear, Biological & Chemical Warfare on the Modern Battlefield. Dulles, VA: Brassey's, Inc., 1997.*
J. Paul, T. V. 2000. *Power versus Prudence: Why Nations Forgo Nuclear Weapons.* Montreal: McGill Queen's University Press.
K. Reiss Mitchell. 1995. *Bridled Ambition: Why Countries Constrain Their Nuclear Capabilities.* Washington, DC: Woodrow Wilson Center.
L. Rhodes, Richard. *The Making of the Atomic Bomb.* Simon and Schuster, New York, (1986 ISBN 978-0-684-81378-3).
M. "Suitcase Nukes"; A Reassessment. Center for Nonproliferation Studies (CNS), September 23, 2002.
N. *WMD: Where Did the Phrase Come from?,* by William Mallon, 2003, History News Network.
O. *The Naked Nuclear Emperor: Debunking Nuclear Deterrence,* Commander Robert Green, Royal Navy (ret), Disarmament and Security Centre, Christchurch, 2000.
P. Sagan, Scott D. and Kenneth N. Waltz. *The Spread of Nuclear Weapons: A Debate Renewed,* 2nd ed. W. W. Norton & Co., 2003.
Q. Tucker, Jonathan, ed., *Toxic Terror: Assessing Terrorist Use of Chemical and Biological Weapons,* MIT Press, Cambridge, MA, 2000.
R. Walter Reich, ed. *Origins of Terrorism: Psychologies, Ideologies, Theologies, States of Mind* (Washington, DC: Woodrow Wilson Center Press, 1998).

Websites and Links: There are hundreds of web resources that deal with WMD. The following list is rich in information on various aspects of WMD, right from types to treaties and from arms control to multilateral disarmament.

A. www.atomicarchive.com features the history, consequences and many more aspects of the atomic age. It contains resources, videos, original texts and documents for the general public, students, policymakers, researchers and educators.
B. www.iiss.org is the website of the International Institute for Strategic Studies (IISS). IISS is a British research institute that focuses on nuclear deterrence, global security, military conflict and arms control.
C. http://library.lanl.gov/infores/nuclear is an information hub on nuclear technologies, nuclear news and reports, nuclear policies, etc.
D. http://epa.gov is the website of the United States Environmental Protection Agency which focuses on enforcing regulations based on Congress laws, and developing better tools and data for the purpose of safeguarding human health and the natural environment upon which life depends.

E. www.acronym.org.uk is the website of the Acronym Institute for Disarmament Diplomacy, which develops and promotes approaches for international security, multilateral disarmament and security agreements.

F. www.fas.org is the website of the Federation of American Scientists (FAS). FAS is an independent, nonpartisan think tank and non-profit membership organization that provides latest, evidence-based news on military doctrines, arms control treaties and many more aspects. It makes policy recommendations on national and international security issues in order to make the world more secure.

G. www.cbrniac.apgea.army.mil is the website of the Chemical, Biological, Radiological and Nuclear Defense Information Analysis Center (CBRNIAC, formerly known as the CBIAC). It is the premier resource for authoritative CBRN defense and homeland security scientific and technical information.

Appendix B

AGENCIES, BODIES AND INSTITUTIONS

Increasing national and international interest in weapons of mass destruction and their global implications has led to the growth of research institutions, private organizations, government bodies and international agencies, which are working relentlessly towards realizing the vision of a world free of WMD. Some focus on implementing strategies to control the spread of WMD, some devote their efforts to reducing or eliminating WMD, some handle issues related to global security and disarmament, and some focus on building a global network to deal with current and future WMD threats through information sharing, research, training, etc. Appendix B contains information and resources on such bodies and agencies.

A. Carnegie Endowment for International Peace (CEIP) is a foreign policy think tank based in Washington, D.C. CEIP is dedicated to advancing cooperation between nations and promoting active international engagement by the United States.

B. Centers for Disease Control and Prevention (CDC) is a US-based federal agency and a global leader in public health, which promotes health and safety through state health departments and other agencies. Its main aim is to provide accurate information and guidelines to enhance the safety and health levels. CDC is a huge resource base on environmental health, data and statistics, emergency preparedness and response, etc. Its core functions are to foster a safe and healthy environment, develop and implement prevention strategies, deal with new health threats, health promotion, etc.

C. National Institutes of Health (NIH) is a medical research agency of the United States Department of Health and Human Services. NIH makes important discoveries and conducts research and training to ensure better safety and health. NIH has a lot of resources on weapons of mass destruction in terms of emergency response and treatment, health effects, etc.

D. Nuclear Threat Initiative (NTI) is a nonpartisan, non-profit organization that focuses on strengthening global security by reducing the risk of use and the spread of nuclear, biological and chemical weapons.

E. Program on International Policy Attitudes (PIPA) is a joint program of the Center for International and Security Studies and the Center on Policy Attitudes. PIPA is devoted to conducting research on the public opinions and attitudes on international matters and foreign policy issues.

F. United Nations Institute for Disarmament Research (UNIDIR) is based in Geneva, Switzerland. UNIDIR is an autonomous institute within the framework of the United Nations. It conducts objective and comprehensive research on both current and future issues related to global security, conflicts and disarmament in order to deal with security and disarmament challenges of today and tomorrow.

G. World Institute for Nuclear Security (WINS) seeks to prevent nuclear terrorism and aims to improve the security of nuclear and high hazard radioactive materials and facilities worldwide. WINS is a premier platform for governments, security experts and international organizations to come together to develop strategies for preventing sabotage, unauthorized access and theft of nuclear materials.

Appendix C

MORE ACADEMIC DEFINITIONS OF WMD

Sr. No.	*Definition*	*Source*
1.	Weapons of mass destruction should be defined to include atomic explosive weapons, radioactive material weapons, lethal chemical and biological weapons, and any weapons developed in the future which have characteristics comparable in destructive effect to those of the atomic bomb or other weapons mentioned above.	UN Committee For Conventional Armaments, August 1948
2.	Weapon of mass destruction/*arme de destruction massive*: A weapon that is capable of a high order of destruction and of being used in such a manner as to destroy people, infrastructure, or other resources on a large scale.	NATO Glossary of Terms and Definitions
3.	Weapons that can cause great damage to very large areas, including nuclear weapons, chemical weapons, and biological weapons. They are often simply referred to as WMD.	Macmillan Dictionary (www.macmillandictionary.com/dictionary/british/weapons-of-mass-destruction)

References

1. American Dialect Society—2002 Words of the Year: http://www.americandialect.org/2002_words_of_the_year
2. "Archbishop's Appeal: Individual Will and Action; Guarding Personality," *The Times* (London), December 28, 1937, 9.
3. "Atmospheric Effects and Societal Consequences of Regional Scale Nuclear Conflicts and Acts of Individual Terrorism," O. B. Toon, R. P. Turco, A. Robock, C. Bardeen, L. Oman and G. L. Stenchikov.
4. Commission for Conventional Armaments, UN document S/C.3/32/Rev.1, August 1948, as quoted in United Nations, Office of Public Information, *The United Nations and Disarmament, 1945–1965*, UN Publication 67.I.8, 28.
5. Commission for Conventional Armaments, *Resolution Adopted by the Commission at Its Thirteenth Meeting, 12 August 1948, and a Second Progress Report of the Commission*, UN document S/C.3/32/Rev.1, 12 August 1948.
6. Definition of Weapons of Mass Destruction in the Macmillan Dictionary: www.macmillandictionary.com/dictionary/british/weapons-of-mass-destruction
7. *Dictionary of Basic Military Terms: A Soviet View*, published under the auspices of the U.S. Air Force, *Soviet Military Thought*, vol. 9, trans. DGIS Multilingual Section, Translation Bureau, Secretary of State Department, Ottawa, Canada (Washington, DC: U.S. Government Printing Office, 1976), 148.
8. Ernest Sternglass, "Infant Mortality Changes Following the Three Mile Island Accident," presented at the 5th World Congress of Engineers and Architects, Tel Aviv, Israel, 1980.
9. In a joint declaration signed by President Harry Truman of the United States, Prime Minister Clement Attlee of the United Kingdom, and Prime Minister Mackenzie King of Canada. See Department of State,

Historical Office, *Documents on Disarmament, 1945–1969, Volume I: 1945–1956*, Publication 7008, August 1960, 1–3.

10. Kofi Annan, We Can Love What We Are, Without Hating What—And Who—We Are Not, UN Press Release SG/SM/8071, October 2001.
11. Morrison, Philip and Kosta Tsipis. 2003. "Rightful Names," *Bulletin of the Atomic Scientists* 59 (May/June): 77.
12. National Cancer Institute, "Questions and Answers on the NCI Fallout Report," pp. 2-3. Released in National Cancer Institute Press Packet, August 1, 1997.
13. "OSHA/EPA Occupational Chemical Database," US Occupational Safety & Health Administration, www.osha.gov
14. *"Rain of Fire: Israel's Unlawful Use of White Phosphorus in Gaza,"* Human Rights Watch Report, Mar 25, 2009.
15. "Ten things you need to know about pandemic influenza," World Health Organization, 14 October 2005.
16. The 1997 National Cancer Institute report, "Estimated Exposures and Thyroid Doses Received by the American People from Iodine-131 in Fallout Following Nevada Atmospheric Nuclear Bomb Tests."
17. Twenty-Ninth Quarterly Report on the Activities of the United Nations Monitoring, Verification and Inspection Commission in Accordance with Paragraph 12 of Security Council Resolution 1284 (1999), UN Security Council document S/2007/314, May 29, 2007.
18. Warwick McKibbin and Alexandra Sidorenko, *Global Macroeconomic Consequences of Pandemic Influenza*, Sydney, Lowy Institute for International Policy, 2006.
19. World Nuclear Stockpile Report 2011—Source: ploughshares.org

Index